P9-DKE-364

THE OXFORD BOOK OF DEATH

THE OXFORD BOOK OF DEATH

CHOSEN AND EDITED BY

D. J. ENRIGHT

OXFORD NEW YORK
OXFORD UNIVERSITY PRESS
1983

Oxford University Press, Walton Street, Oxford OX2 6DP
London Glasgow New York Toronto
Delhi Bombay Calcutta Madras Karachi
Kuala Lumpur Singapore Hong Kong Tokyo
Nairobi Dar es Salaam Cape Town
Melbourne Auckland
and associated companies in
Beirut Berlin Ibadan Mexico City Nicosia

Oxford is a trade mark of Oxford University Press

British Library Cataloguing in Publication Data
The Oxford book of death.
1. Death
I. Enright, D. J.
308.8′8 BD444
ISBN 0-19-214129-5

Manufactured in the United States of America

CONTENTS

EDITOR'S NOTE

DATES of authors or of publications are given on their first appearance. In the case of translated items the translator is named in the first reference to a particular work; when no translator is mentioned, the editor is himself responsible for the English version.

The editor wishes to thank the following for help of various kinds: Jonathan Barker (The Arts Council Poetry Library), Amanda Beresford, Leslie Booth, Michael Booth (The London Library), Hugo Brunner, Shirley Chew, Frank Cioffi, August Closs, Terrence Des Pres (*The Survivor*, 1976), Madeleine Enright, Bryan Healing, Karin Horowitz, Tony Kitzinger, Roger Pringle (*Poems of Warwickshire*, 1980), David Rawlinson, John Silverlight, Jacqueline Simms, G. Singh, Norah Smallwood, Jane Turner, Patricia Utechin (*Epitaphs from Oxfordshire*, 1980), and Margot Walmsley.

D. J. E.

The human race is the only one that knows it must die, and it knows this only through its experience. A child brought up alone and transported to a desert island would have no more idea of death than a cat or a plant.

VOLTAIRE

Who the living would explain
He must enter death's domain.

CHRISTIAN MORGENSTERN

In my happier days I used to remark on the aptitude of the saying, 'When in life we are in the midst of death.' I have since learnt that it's more apt to say, 'When in death we are in the midst of life.'

A BELSEN SURVIVOR

And 'the earth under our feet,'
we say glibly, hating
the 'Esoteric' which is not
to be included in our anthologies, the
unthinkable . . .

WILLIAM CARLOS WILLIAMS

Introduction

THE first words in a collection of last words entitled *The Art of Dying* (1930) are these:

> For what was there none cared a jot;
> But all were wroth with what was not.

More than is generally the case with compilations, by its nature the present one is bound to be incomplete. Its subject is exceptionally large and of exceptionally common concern. In this line of country the compiler could easily and even pleasurably persist in his task until death should part the one from the other.

One of the features not invariably to be found here is certainty: the country in question is one from which no indisputably trustworthy traveller has returned to tell the tale. But the uses of certainty are themselves uncertain in this sphere, and what, I suspect, is most sadly missing is music, the art which seemingly is able to tell us more about heaven, and rival literature's eloquence on the subject of hell.

That man is the only animal who is conscious (from time to time) that it must die is a truism. Possibly it is less banal to observe that in every age man has written about death, its nature, its forms, processes and implications, and written bravely, unsparingly, gracefully and wittily. Reading for this anthology, I was moved to the thought that on no theme have writers shown themselves more lively. Nor is this some professional peculiarity: writers are in no fundamental respect different from other members of the race to which they belong. The proposition 'life eminent creates the shade of death' can be turned around with equal truth. Solicitous friends—if I may be permitted a personal note—feared that the 'depressing' nature of the undertaking might prove too much for the compiler's animal spirits. The reverse was the case. Or most of the time, for it must be admitted that certain areas—suicide, the death of children—called for more fortitude than others.

Another source of distress was 'documentary' material, the personal utterances, straight out of immediate suffering, in which rawness of feeling is untempered by the skills and benefits of art. Not surprisingly, for the most part the best-written pieces had most to say, but—this being a subject on which there are no real experts—lay voices rightly insisted on being heard alongside those of 'literary men'. An extra sadness came from noting the dates of those cited, the short lives of so many of them—Marlowe, Keats, Shelley, Emily Brontë, Corbière, Rosenberg, Owen, Keith Douglas, Tadeusz Borowski, to mention only some who did not outlive their thirtieth year.

Though convenient, the division of the book into fourteen sections is to some extent artificial. Definitions and 'justifications' of death are hardly to be distinguished from attitudes to death; Heavens and Hells need Immortality, and quite possibly vice versa; Proust's account of Bergotte's dying could belong as aptly to 'The Hour of Death' as to 'Resurrections and Immortalities'; suicide reappears in the section devoted to Love and even in the section on Children; grief manifests itself almost everywhere, as also perhaps, in one shape or another, does consolation. Even so, I hope that some sense will be found in the arrangement of material and the occasional interrelations. To press a little further—if the reader detects traces of a pattern in the book, then those traces were either intended or accepted with gratitude; if he does not, then how could a sprawling miscellany of views, surmises and responses, drawn from the greater part of recorded time and from much of known space, ever be expected to show a pattern?

The great initial uncertainty had to do with the public status of the subject. Was death a popular, much-canvassed topic, the 'in thing' or at least (as someone proposed more accurately) the 'coming thing'? Or was it the very opposite—a dirty little secret, a thing of shame, the last taboo in an otherwise totally uninhibited world? The newish breed of psychologists or social workers called 'thanatologists' is of the latter opinion. In 1975 Elisabeth Kübler-Ross described death as 'a dreaded and unspeakable issue to be avoided by every means possible in our modern society'. Back in 1940 J. C. Flugel remarked on the similarity between attitudes to death and attitudes to sex and the parallel reactions of flight, repression and symbolism. 'We talk of a "departed" friend as we talk of a "fallen" woman, shunning in both cases the greater affect that a more direct expression might involve . . .' Euphemisms make an interesting and by no means elementary study, but true, much seems to have happened since the days when we used to talk of fallen women, and in 1955 Geoffrey Gorer spoke of the 'shift of prudery' whereby copulation had become mentionable and natural death (as opposed to violent death in films, books and horror comics) increasingly unmentionable.

More recently, however, a writer in *The Observer* (1980) contrived to have it both ways: 'Death is the media's new radical chic subject, fodder for endless TV programmes and newspaper articles. Bereavement, euthanasia, hospices—all have come out of the closet and into the limelight. So can we still be a death-denying society? Though death may be the subject of intellectual debate, it is our most pervasive daily taboo.' Once again, it would appear from this account, intellectuals are to be found noisily haunting their taboo-free ivory tower.

What is one to believe? In the course of my task I was led to the conclusion that there is no more of a 'taboo' on death than there is on sex, although for obvious reasons the former is less frequently appealed to and less heavily exploited. (Only in warnings against the use of tobacco are the two conjoined: smokers are never kissed, and smokers die.) While to 'deny' death would sound as foolish as the lady who told Carlyle she had decided to accept the universe, I cannot say that I share the thanatologists' missionary urge to bring death out into the open. Much of the time, as a subject of conversation or even of solitary thought, it belongs elsewhere—standing to one side of life, which being so much shorter deserves to be given priority. It is another matter entirely to sympathize with what John McManners has recently written: 'The attitude of men to the death of their fellows is of unique significance for an understanding of our human condition . . . The knowledge that we must die gives us our perspective for living, our sense of finitude, our conviction of the value of every moment, our determination to live in such a fashion that we transcend our tragic limitation.' The 'human condition' is larger than the condition of the death-bed, and of longer standing.

The passage quoted above implies a justification for compilations of the present kind; it certainly describes the intention which, albeit less consciously, moved the present compiler. In *The Hour of Our Death* Philippe Ariès observes that 'serious works on death are never completely free of ambiguity'. While the remark applied specifically to the way love or lust finds itself embraced in the subject, I take it, more generally, to indicate how such works quickly branch out into multifarious highways and byways. It is not so much that I have been indulgent in interpreting my terms of reference as that the theme forced liberality upon me. Life helps us to shape our thoughts about death, and often serves as our metaphor, the known invoked to adumbrate the as yet unexperienced. Hence to talk at all interestingly about death is inevitably to talk about life.

Definitions

DEFINE death? But does the subject really exist? Seemingly not, or hardly at all. 'Death is not lived through,' said Wittgenstein. And according to Freud, 'Death is an abstract concept with a negative content for which no unconscious correlative can be found'—a ruling which (even if one cannot be sure of understanding it) clearly puts the subject in its non-place. The scientific answers please by their neatness more than satisfy by the light they cast. You know what life is—well, death is the cessation or absence of it. It is the price paid for virtually all of the properties we value in life. (An old and persistent theory has it that every ejaculation of semen takes a day off a man's life: no wonder women live longer.) If there were no life about, there would be no death. So you know where to lay the blame.

Rather more rewarding is Freud's explanation of the phenomenon in the passage from *Beyond the Pleasure Principle* quoted in the following pages. Death (in the form of non-existence) preceded life, and evolution has been a continuous struggle against back-sliding into the *status quo ante*; or: life is the longest distance or most roundabout route we can contrive between the two points of birth and death. So at least we find our own paths to death, circumstances permitting.

Much of this argument is summed up elegantly and lucidly, with the help of a homely metaphor, in Francis Quarles's quatrain, 'He that begins to live, begins to die.' It would appear that literary authors have the edge over scientific writers, though this dichotomy of activities did not exist among the ancients, for whom all writers (apart perhaps from love-lyrists) were 'philosophers'. Poets can be as precise as scientists both on the point that when we die, we do not exist, so death doesn't exist for us, and as concerns the wider and more positive compensations attached. I am not thinking only of the comforts of Christianity, as when, in an ingenious and shop-talking comparison, Donne likens death to translation into 'a better language' (as opposed to remaindering) and accession into the great library of eternity, or St Francis more simply blesses God 'for our sister, the death of the body'. For while Darwin talks unappealingly about the consequent 'production of the higher animals', Blake even derives consolation from the thought of worms nourished by his 'shining woman'. And Montaigne, an old-style philosopher, justifies the ways of death economically, without the aid of terminology from industry or the battlefield: 'Make room for others, as others have done for you.'

In the portrayal of Death anthropomorphized ('the ruffian on the stairs', 'the Rector's pallid neighbour at The Firs'), literary people, being traditionally excitable and irresponsible, have an advantage in that they are permitted or even expected to speculate freely and 'make up' things. In plotting the topography of an undiscovered country, in this case embracing the provinces of heaven and hell, speculation is the only alternative to inertia and silence. We may not be too eager to bring death 'into the limelight', but we cannot cut it dead. What Dr Johnson said about 'delusion' in the theatre can be applied here: speculation, if speculation be admitted, has no certain limitation. Yet we can still distinguish between the tellingly imaginative and the weakly or wishfully fanciful, between shared exploration and private obsession.

Who telleth a tale of unspeaking death?
  Who lifteth the veil of what is to come?
Who painteth the shadows that are beneath
  The wide-winding caves of the peopled tomb?
Or uniteth the hopes of what shall be
With the fears and the love for that which we see?

PERCY BYSSHE SHELLEY (1792–1822), from 'On Death'

When compared with the stretch of time unknown to us, O king, the present life of men on earth is like the flight of a single sparrow through the hall where, in winter, you sit with your captains and ministers. Entering at one door and leaving by another, while it is inside it is untouched by the wintry storm; but this brief interval of calm is over in a moment, and it returns to the winter whence it came, vanishing from your sight. Man's life is similar; and of what follows it, or what went before, we are utterly ignorant.

THE VENERABLE BEDE (*c.*673–735), *Ecclesiastical History of the English People*

We're faced, aren't we, with an initial question of the utmost significance—the question, namely, of whether our subject can be said, in any meaningful sense, to exist at all . . . I have been accused of arrogance on the grounds that I have described certain areas of previous philosophical debates as areas of literal nonsense. I still hold to that description, though I have modified certain of my strictures in directions which are probably rather too technical for discussion here. But I wouldn't wish it to be thought that this act of definition puts a full-stop, as it were, to any further discussion. The question of why metaphysics, to take the most obvious example, seems to me to be a subject lacking in intelligible content is still a question for debate. I mean that this view of metaphysical speculation has not yet been as widely accepted as I would presumably wish it to be. I say 'presumably' because I recognize some justice in my friend Rexforth's often-repeated gibe that a complete conversion of the world to my views would leave me without the only occupation which I really enjoy. What I'm trying to make clear is that I shall be happy to discuss, at almost any length, the question of whether death may properly be regarded as a subject for discussion.

MAURICE RICHARDSON and PHILIP TOYNBEE, *Thanatos: A Modern Symposium*, 1963 (the philosopher Strip is speaking)

If we are aware of what indicates life, which everyone may be supposed to know, though perhaps no one can say that he truly and clearly understands what constitutes it, we at once arrive at the discrimination of death. It is the cessation of the phenomena with which we are so especially familiar—the phenomena of life.

J. G. SMITH, *Principles of Forensic Medicine*, 1821

Death is the price paid by life for an enhancement of the complexity of a live organism's structure.

ARNOLD TOYNBEE (1889–1975), *Life After Death*

The attributes of life were at some time evoked in inanimate matter by the action of a force of whose nature we can form no conception. It may perhaps have been a process similar in type to that which later caused the development of consciousness in a particular stratum of living matter. The tension which then arose in what had hitherto been an inanimate substance endeavoured to cancel itself out. In this way the first instinct came into being: the instinct to return to the inanimate state. It was still an easy matter at that time for a living substance to die; the course of its life was probably only a brief one, whose direction was determined by the chemical structure of the young life. For a long time, perhaps, living substance was thus being constantly created afresh and easily dying, till decisive external influences altered in such a way as to oblige the still surviving substance to diverge ever more widely from its original course of life and to make ever more complicated *détours* before reaching its aim of death. These circuitous paths to death, faithfully kept to by the conservative instincts, would thus present us today with the picture of the phenomena of life. If we firmly maintain the exclusively conservative nature of instincts, we cannot arrive at any other notions as to the origin and aim of life.

The implications in regard to the great group of instincts which, as we believe, lie behind the phenomena of life in organisms must appear no less bewildering. The hypothesis of self-preservative instincts, such as we attribute to all living beings, stands in marked opposition to the idea that instinctual life as a whole serves to bring about death. Seen in this light, the theoretical importance of the instincts of self-preservation, of self-assertion and of mastery greatly diminishes. They are component instincts whose function it is to assure that the organism shall follow its own path to death, and to ward off any possible ways of returning to inorganic existence other than those which are immanent in the organism itself. We have no longer to reckon with the organism's puzzling determination (so hard to fit into any context) to maintain its own existence in the face of every obstacle. What we are left with is the fact that the organism wishes to die only in its own fashion. Thus these guardians of life, too, were originally the myrmidons of death. Hence arises the paradoxical situation that the living organism struggles most energetically against events (dangers, in fact) which might help it to attain its life's aim rapidly—by a kind of short-circuit.

SIGMUND FREUD (1856–1939), *Beyond the Pleasure Principle*, tr. James Strachey (a later footnote asks the reader not to overlook the fact that 'what follows is the development of an extreme line of thought')

AS. SOONE. AS. WEE. TO. BEE. BEGVNNE:
WE. DID. BEGINNE. TO. BE. VNDONE.

English *memento mori* medal, *c.*1650

Forbear, fond taper: what thou seek'st, is fire:
Thy own destruction's lodg'd in thy desire,
Thy wants are far more safe than their supply:
He that begins to live, begins to die.

FRANCIS QUARLES (1592–1644), *Hieroglyphics of the Life of Man*

Thus, from the war of nature, from famine and death, the most exalted object which we are capable of conceiving, namely, the production of the higher animals, directly follows. There is grandeur in the view of life, with its several powers, having been originally breathed into a few forms or into one; and that, while this planet has gone cycling on according to the fixed laws of gravity, from so simple a beginning endless forms most beautiful and most wonderful have been, and are being, evolved.

CHARLES DARWIN (1809–82), *On the Origin of Species*

Thanatos. Also, death instinct. One of two primal instincts attributed to man, the other being Eros. The function of Thanatos is to restore higher organic organization to a simpler, pre-vital state. In this, Thanatos is expressing a tendency evident elsewhere in nature for organization to run down into greater simplicity. Thus this instinct tends to reinstate earlier levels of development by impelling the individual towards passivity, and to bring about the cessation of vital integrity in the organism through injury and destruction. Human life proceeds as a compromise between the two primal instincts. The compromise is sometimes poorly made, with the result that human life and products are squandered and destroyed. When the compromise is well made, nations can flourish in power and beauty. The laws that regulate the compromise are obscure.

DONALD M. KAPLAN and ARMAND SCHWERNER, *The Domesday Dictionary*, 1963

The power of population is so superior to the power in the earth to produce subsistence for man, that premature death must in some shape or other visit the human race. The vices of mankind are active and able ministers of depopulation. They are the precursors in the great army of destruction; and often finish the dreadful work themselves. But should they fail in this war of extermination, sickly seasons, epidemics, pestilence and plague advance in terrific array, and sweep off their thousands and ten thousands. Should success be still incomplete, gigantic inevitable famine stalks in the rear, and with one mighty blow levels the population with the food of the world.

THOMAS ROBERT MALTHUS (1766–1834), *An Essay on the Principle of Population*

Every extraordinary person has a particular mission which he is called upon to fulfil. When he has accomplished it, he is no longer needed on earth in the same form, and Providence uses him for something else . . . Mozart died at thirty-six. Raphael at practically the same age. Byron was only a little older. But each of them had accomplished his mission perfectly, and it was time for them to go so that others might still have something left to do in a world created to last a long while.

JOHANN WOLFGANG VON GOETHE (1749–1832), *Conversations with Eckermann*, 1828

O death, we thank you for the light you cast on our ignorance: you alone convince us of our lowliness, you alone make us know our dignity . . . All things summon us to death: nature, almost envious of the good she has given us, tells us often and gives us notice that she cannot for long allow us that scrap of matter she has lent . . . she has need of it for other forms, she claims it back for other works.

JACQUES-BÉNIGNE BOSSUET (1627–1704), *Sermons. On Death*

Look round our World; behold the chain of Love
Combining all below and all above.
See plastic Nature working to this end,
The single atoms each to other tend,
Attract, attracted to, the next in place
Form'd and impell'd its neighbour to embrace.
See Matter next, with various life endu'd,
Press to one centre still, the gen'ral Good.
See dying vegetables life sustain,
See life dissolving vegetate again:
All forms that perish other forms supply,
(By turns we catch the vital breath, and die)
Like bubbles on the sea of Matter born,
They rise, they break, and to that sea return.

ALEXANDER POPE (1688–1744), *An Essay on Man*

'O little Cloud,' the virgin said, 'I charge thee tell to me
Why thou complainest not, when in one hour thou fad'st away:
Then we shall seek thee, but not find. Ah! Thel is like to thee—
I pass away; yet I complain, and no one hears my voice.'

'. . . O maid, I tell thee, when I pass away,
It is to tenfold life, to love, to peace, and raptures holy.
Unseen descending, weigh my light wings upon balmy flowers,
And court the fair-eyed dew to take me to her shining tent:
The weeping virgin, trembling, kneels before the risen sun,
Till we arise, linked in a golden band, and never part,
But walk united, bearing food to all our tender flowers.'

'Dost thou, O little Cloud? I fear that I am not like thee,
For I walk through the vales of Har, and smell the sweetest flowers,
But I feed not the little flowers. I hear the warbling birds,
But I feed not the warbling birds: they fly and seek their food.
But Thel delights in these no more, because I fade away,
And all shall say: "Without a use this shining woman lived;
Or did she only live to be at death the food of worms?" '

The Cloud reclined upon his airy throne, and answered thus:
'Then if thou art the food of worms, O virgin of the skies,
How great thy use, how great thy blessing! Everything that lives
Lives not alone, nor for itself . . .'

WILLIAM BLAKE (1757–1827), *The Book of Thel*

If we were to live here always, with no other care than how to feed, clothe, and house ourselves, life would be a very sorry business. It is immeasurably heightened by the solemnity of death. The brutes die even as we; but it is our knowledge that we have to die which makes us human. If nature cunningly hides death, and so permits us to play out our little games, it is easily seen that our knowing it to be inevitable, that to every one of us it will come one day or another, is a wonderful spur to action. We really do work while it is called today, because the night cometh when no man can work . . . And knowing that his existence here is limited, a man's workings have reference to others rather than to himself, and thereby into his nature comes a new influx of nobility. If a man plants a tree, he knows that other hands than his will gather the fruit; and when he plants it, he thinks quite as much of those other hands as of his own. Thus to the poet there is the dearer life after life; and posterity's single laurel leaf is valued more than a multitude of contemporary bays. Even the man immersed in money-making does not make money so much for himself as for those who may come after him.

ALEXANDER SMITH (1830–67), *Dreamthorp*

The somewhat greater size and strength of males leads most people to think that males are biologically superior; but just the opposite is true. In the modern nations and in the cities of developing nations, males throughout their lives are more likely to die than are females of the same age. More than 125 males are conceived for every 100 females, but the proportion of males born alive is much less than that; in the United States and Britain, the figure is about 106 males to 100 females. Several causes are behind the disproportionately high loss of males between conception and birth: the greater number of male foetuses that are spontaneously aborted, the greater likelihood that a male will die from birth trauma, and the unusually large number of males who suffer from congenital abnormalities. Males continue to die at a greater rate throughout their lives. During the first year of life, about fifty-four males die for every forty-six females; by age twenty-one, the figure is about sixty-eight males for every thirty-two females of the same age. Decade after decade male mortality rises—until about age seventy-five, when the proportions are reversed, simply because so few males older than that are still living . . .

No simple explanation for the disproportionately high number of male deaths has yet emerged. Part of the explanation lies in the preponderance among males of certain genetic disorders, but that does not account for the high death rate of males from diseases to which both sexes are susceptible. Epilepsy, for example, attacks males and females in approximately equal numbers, but the death rate from it is about 30 per cent higher in males. And females suffering from the same infectious diseases as males die at a much lower rate. A comparison of groups of males and females who smoked equally large numbers of cigarettes showed that females generally were more resistant than males to such deleterious effects as lung cancer and heart disease. Females obviously possess some superior capacity for survival that has little to do with the kinds of lives they lead.

PETER FARB, *Humankind*, 1978

So death, the most terrifying of ills, is nothing to us, since so long as we exist, death is not with us; but when death comes, then we do not exist. It does not then concern either the living or the dead, since for the former it is not, and the latter are no more.

EPICURUS (341–270 BC), *Letter to Menoeceus*, tr. Cyril Bailey

Nothing that is extreme is evil. Death comes to you? It would be dreadful could it remain with you; but of necessity either it does not arrive or else it departs.

SENECA (4 BC–AD 65), *Letters to Lucilius*

Death is not an event of life. Death is not lived through. If by eternity is understood not endless temporal duration but timelessness, then he lives eternally who lives in the present. Our life is endless in the way that our visual field is without limit. The temporal immortality of the human soul, that is, its eternal survival after death, is not only in no way guaranteed, but this assumption will not do for us what we have always tried to make it do. Is the riddle solved by the fact that I survive forever? Is this eternal life not as enigmatic as our present one?

LUDWIG WITTGENSTEIN (1889–1951), *Tractatus Logico-Philosophicus*, tr. C. K. Ogden

The World a Hunting is,
The prey, poor Man, the Nimrod fierce is Death,
His speedy Greyhounds are
Lust, Sickness, Envy, Care,
Strife that ne'er falls amiss,
With all those ills which haunt us while we breathe.
Now if (by chance) we fly
Of these the eager Chase,
Old Age with stealing Pace
Casts up his Nets, and there we panting die.

WILLIAM DRUMMOND OF HAWTHORNDEN (1585–1649)

Alone of gods Death has no love for gifts,
Libation helps you not, nor sacrifice.
He has no altar, and he hears no hymns;
From him alone Persuasion stands apart.

AESCHYLUS (525–456 BC), *Niobe*, tr. C. M. Bowra

Hold! Pale Death, at the poor man's shack and the pasha's palace kicking
Impartially, announces his arrival.
Life's brief tenure forbids high hopes to be built in disproportion,
My lucky Sestius, for Night and Pluto's
Shadowy walls and the ghosts men talk of will soon be crowding round you.
Once there, you cannot rule the feast by dice-throw
Or give Lycidas long rapt gazes.

HORACE (65–8 BC), *Odes*, tr. James Michie

Think, in this batter'd Caravanserai
Whose Doorways are alternate Night and Day,
How Sultan after Sultan with his Pomp
Abode his Hour or two, and went his way.

They say the Lion and the Lizard keep
The Courts where Jamshyd gloried and drank deep:
  And Bahram, that great Hunter—the Wild Ass
Stamps o'er his Head, and he lies fast asleep.

OMAR KHAYYAM (*c.*1050–*c.*1123), in Edward Fitzgerald's version, *The Rubaiyat*, 1859

O sons of men,
Lean death perches upon your shoulder
Looking down into your cup of wine,
Looking down on the breasts of your lady.
You are caught in the web of the world
And the spider Nothing waits behind it.
Where are the men with towering hopes?
They have changed places with owls,
Owls who lived in tombs
And now inhabit a palace.

The Thousand and One Nights, 10th century onwards, tr. E. Powys Mathers

The boast of heraldry, the pomp of pow'r,
  And all that beauty, all that wealth e'er gave,
Await alike th'inevitable hour:
  The paths of glory lead but to the grave.

Nor you, ye proud, impute to these the fault
  If Memory o'er their tomb no trophies raise,
Where through the long-drawn aisle and fretted vault
  The pealing anthem swells the note of praise.

Can storied urn or animated bust
  Back to its mansion call the fleeting breath?
Can Honour's voice provoke the silent dust,
  Or Flatt'ry soothe the dull cold ear of death?

THOMAS GRAY (1716–71), from 'Elegy written in a Country Church Yard'

'Oh, father Utnapishtim, you who have entered the assembly of the gods, I wish to question you concerning the living and the dead, how shall I find the life for which I am searching?'

Utnapishtim said, 'There is no permanence. Do we build a house to stand for ever, do we seal a contract to hold for all time? Do brothers divide an inheritance to keep for ever, does the flood-time of rivers endure? It is only the nymph of the dragonfly who sheds her larva and sees the sun in his glory. From the days of old there is no permanence. The sleeping and the dead, how alike they are, they are like a painted death. What is there between the master and the servant when both have fulfilled their doom? When the Anunnaki, the judges, come together, and Mammetun the mother of destinies, together they decree the fates of men. Life and death they allot, but the day of death they do not disclose.'

The Epic of Gilgamesh, ?2000 BC, tr. N. K. Sandars

A young fir-tree grows
Who knows where, in what forest,
A rosebush, in which garden
Who can tell?
Already chosen
(Soul, remember this!)
To strike root on your grave
And flourish,

Two young black horses graze
Along the lea.
Townwards returning,
They curvet merrily.
Step by step they'll pace,
Your corpse attending—
Before their hooves can lose,
It may be, may be,
The iron shoes
Whose quick sparks I see.

EDUARD MÖRIKE (1804–75), 'Soul, remember this!'

Truly do we live on earth?
Not forever on earth; only a little while here.
Although it be jade, it will be broken,
Although it be gold, it is crushed,
Although it be quetzal feather, it is torn asunder.
Not forever on earth; only a little while here.

Nahuatl text in *Colección de Cantares Mexicanos*, *c.*1560, tr. Jack Emory Davis

The glories of our blood and state
  Are shadows, not substantial things;
There is no armour against Fate,
  Death lays his icy hand on kings;
    Sceptre and crown
    Must tumble down,
And in the dust be equal made,
With the poor crooked scythe and spade.

Some men with swords may reap the field,
  And plant fresh laurels where they kill;
But their strong nerves at last must yield,
  They tame but one another still;
    Early or late,
    They stoop to Fate,
And must give up the murmuring breath,
When they, pale captives, creep to Death.

The garlands wither on your brow,
  Then boast no more your mighty deeds;
Upon Death's purple altar now,
  See where the victor-victim bleeds;
    Your heads must come,
    To the cold tomb;
Only the actions of the just
Smell sweet, and blossom in their dust.

James Shirley (1596–1666), *The Contention of Ajax and Ulysses*

Death must simply become the discreet but dignified exit of a peaceful person from a helpful society that is not torn, not even overly upset by the idea of a biological transition without significance, without pain or suffering, and ultimately without fear.

Philippe Ariès, *The Hour of Our Death*, 1977, tr. Helen Weaver

The death of a man is like the fall of a mighty nation
That had valiant armies, captains and prophets,
And wealthy ports and ships over all the seas,
But now it will not relieve any besieged city,
It will not enter into any alliance,
Because its cities are empty, its population dispersed,
Its land once bringing harvest is overgrown with thistles,
Its mission forgotten, its language lost,
The dialect of a village high upon inaccessible mountains.

Czesław Miłosz (b. 1911), 'The Fall', tr. the author and Lillian Vallee

When a man dies
his portraits change.
His eyes look in a different way,
his lips smile a different smile.
I noticed this on returning
from the funeral of a poet.
Since then I have often checked it,
and my theory has been confirmed.

ANNA AKHMATOVA (1889–1966), tr. Richard McKane

All mankind is of one author, and is one volume; when one man dies, one chapter is not torn out of the book, but translated into a better language; and every chapter must be so translated; God employs several translators; some pieces are translated by age, some by sickness, some by war, some by justice; but God's hand is in every translation, and his hand shall bind up all our scattered leaves again for that library where every book shall lie open to one another.

JOHN DONNE (1572–1631), *Devotions*

'Death,' said Mark Staithes. 'It's the only thing we haven't succeeded in completely vulgarizing. Not from any lack of the desire to do so, of course. We're like dogs on an acropolis. Trotting round with inexhaustible bladders and only too anxious to lift a leg against every statue. And mostly we succeed. Art, religion, heroism, love—we've left our visiting-card on all of them. But death—death remains out of reach. We haven't been able to defile *that* statue. Not yet, at any rate. But progress is still progressing.'

ALDOUS HUXLEY (1894–1963), *Eyeless in Gaza*

After all, what *is* death? Just nature's way of telling us to slow down.

DICK SHARPLES, *In Loving Memory*, Yorkshire Television, 1979 (according to A. Alvarez, an American insurance proverb)

Who was it that said the living are the dead on holiday?

TERRY NATION, *Dr Who: Destiny of the Daleks*, BBC TV, 1980

When a man sleeps, often his dream will break
With too much dreamed emotion, dream's excess;
So from the dream of life I now may wake,
Cloyed with emotion, to death's wakefulness.

HUGO VON HOFMANNSTHAL (1874–1929), *Death and the Fool*, tr. Michael Hamburger

Pensive here I sat
Alone, but long I sat not, till my womb,
Pregnant by thee, and now excessive grown,
Prodigious motion felt and rueful throes.
At last this odious offspring whom thou seest,
Thine own begotten, breaking violent way
Tore through my entrails, that with fear and pain
Distorted, all my nether shape thus grew
Transformed; but he my inbred enemy
Forth issued, brandishing his fatal dart
Made to destroy. I fled, and cried out *Death!*
Hell trembled at the hideous name, and sighed
From all her caves, and back resounded *Death!*
I fled, but he pursued (though more, it seems,
Inflamed with lust than rage) and swifter far,
Me overtook, his mother, all dismayed,
And in embraces forcible and foul
Engend'ring with me, of that rape begot
These yelling monsters that with ceaseless cry
Surround me, as thou saw'st, hourly conceived
And hourly born, with sorrow infinite
To me; for when they list, into the womb
That bred them they return, and howl and gnaw
My bowels, their repast; then bursting forth
Afresh with conscious terrors vex me round,
That rest or intermission none I find.
Before mine eyes in opposition sits
Grim Death my son and foe, who sets them on,
And me his parent would full soon devour
For want of other prey, but that he knows
His end with mine involved . . .

JOHN MILTON (1608–74), *Paradise Lost* (Sin addressing Satan)

At first no one needed to die. Then one day God wanted to see whether man or snake was worthy of immortality, and so he arranged a race between them. During the race the man met a woman. He stopped to smoke and chat with her for so long that the snake reached God first. And so God told man: The snake is worthier than you—it shall be immortal, but you shall die, and all your race.

African story, Gudji and Darasa tribes

The last deed of Maui the Trickster, the Polynesian demigod who played jokes, pushed the sky higher, roped the sun with braided pubic hair from his mother, pulled the land up out of the ocean, and brought fire to the earth, was to seek immortality for men and women by stealing it from Hina of the Night. He instructed the people, the beasts, the birds, and the elements to be silent. Hunters walked through forests and fishermen waited in this same silence. In silence the snarer caught birds alive, plucked the few red feathers, and released them; the seer read the clouds, heard spirits and did not disturb them. Children learned and worked silently. There was a chant that could hardly be discerned from silence. Maui dived into the ocean, where he found great Hina asleep. Through her vagina like a door, he entered her body. He took her heart in his arms. He had started tunnelling out feet first when a bird, at the sight of his legs wiggling out of the vagina, laughed. Hina awoke and shut herself, and Maui died.

MAXINE HONG KINGSTON (b. 1940), *China Men*

Lo! 'tis a gala night
  Within the lonesome latter years!
An angel throng, bewinged, bedight
  In veils, and drowned in tears,
Sit in a theatre, to see
  A play of hopes and fears,
While the orchestra breathes fitfully
  The music of the spheres.

Mimes, in the form of God on high,
  Mutter and mumble low,
And hither and thither fly—
  Mere puppets they, who come and go
At bidding of vast formless things
  That shift the scenery to and fro,
Flapping from out their Condor wings
  Invisible Woe!

That motley drama—oh, be sure
  It shall not be forgot!
With its Phantom chased for evermore
  By a crowd that seize it not,
Through a circle that ever returneth in
  To the self-same spot,
And much of Madness, and more of Sin,
  And Horror the soul of the plot.

But see, amid the mimic rout,
A crawling shape intrude!
A blood-red thing that writhes from out
The scenic solitude!
It writhes!—it writhes!—with mortal pangs
The mimes become its food,
And seraphs sob at vermin fangs
In human gore imbued.

Out—out are the lights—out all!
And, over each quivering form,
The curtain, a funeral pall,
Comes down with the rush of a storm,
While the angels, all pallid and wan,
Uprising, unveiling, affirm
That the play is the tragedy, 'Man',
And its hero, the Conqueror Worm.

EDGAR ALLAN POE (1809–49), 'The Conqueror Worm'

The Rector's pallid neighbour at The Firs,
Death, did not flurry the parishioners.
Yet from a weight of superstitious fears
Each tried to lengthen his own term of years.
He was congratulated who combined
Toughness of flesh and weakness of the mind
In consequential rosiness of face.
This dull and not ill-mannered populace
Pulled off their caps to Death, as they slouched by,
But rumoured him both atheist and spy.
All vowed to outlast him (though none ever did)
And hear the earth drum on his coffin-lid.
Their groans and whispers down the village street
Soon soured his nature, which was never sweet.

ROBERT GRAVES (b. 1895), 'The Villagers and Death'

I picture Death as being fairly tall, having clearly cut features. His nose is straight and his eyes sparkle. He is a little on the slim side though looks fairly powerful. He has a very dark goatee coming to a point, black hair which is in need of a haircut. He is wearing a dark suit. He also has a light complexion. I see Death right now almost in the same way as the devil.

Death is a quiet person and kind of scary. He does not say much but is very sharp and it is almost impossible to outsmart him. Although you would like to stay away from him, there's something about him that kind of draws you to him. You like him and fear him at the same time. I picture Death as being millions of years old but only looking about 40.

AMERICAN UNDERGRADUATE, male, quoted in *The Psychology of Death*, ed. Robert Kastenbaum and Ruth Aisenberg, 1972

Death has got something to be said for it:
There's no need to get out of bed for it;
Wherever you may be,
They bring it to you, free.

KINGSLEY AMIS (b. 1922), 'Delivery Guaranteed'

Death began to hammer on the door. HOO HOO HOO! it yelled. LET ME IN!

Go away, said Kleinzeit. Not your time yet.

HOO HOO! yelled Death. I'LL BLOODY TEAR YOU APART. ANY TIME'S MY TIME, I WANT YOU NOW AND I'M GOING TO HAVE YOU NOW. NOW NOW NOW.

Kleinzeit went to the door, double-locked it, fastened the chain. Go away, he said. You're not real, you're just in my mind.

IS YOUR MIND REAL? said Death.

Of course my mind's real, said Kleinzeit.

THEN SO AM I, said Death. THERE I HAVE YOU, EH? It stuck its fingers through the letter box. Bristling black and hairy, with disgusting-looking long grey fingernails.

Kleinzeit grabbed the frying-pan from the kitchen, slammed the hairy black fingers with all his strength.

I'LL GET YOU LATER, said Death, SEE IF I DON'T.

Right, said Kleinzeit. He went back to the plain deal table to start the second paragraph.

When Kleinzeit opened the door of his flat Death was there, black and hairy and ugly, no bigger than a medium-sized chimpanzee with dirty fingernails.

Not all that big, are you, said Kleinzeit.

Not one of my big days, said Death. Sometimes I'm tremendous.

Kleinzeit trotted off down the street. Not too much at first, he reminded himself. Just from here to Thomas More, then fifty steps of walking.

Death followed him chimpanzee-style, putting its knuckles down on the pavement and swinging its legs forward. You're pretty slow, it said.

With glue my heart is laden, said Kleinzeit.

What do you mean? said Death, moving up beside him.

I mean life is gluey, said Kleinzeit. Everything's all stuck together. That isn't what I mean. Everything is *un*stuck, runs over into everything else. Clocks and sparrows, harrows, flocks and crocks, green turtles, Golden Virginia. Yellow paper, foolscap, Rizla. Is there an existence that is only mine?

What's the difference if there is or not? said Death. Does it matter?

You're very friendly, very cosy, very matey today, said Kleinzeit. How do I know you won't start yelling HOO HOO again and come at me all of a sudden?

You don't, said Death. But right now I feel friendly. It's lonely for me, you know. Lots of people think I'm beastly.

Kleinzeit looked down at Death's black bristly back rising and falling as it swung along beside him. You are, you know, he said.

Death looked up at him, wrinkled back its chimpanzee lips, showed its yellow teeth. Be nice, it said. One day you'll need me.

Under the bed Death sat humming to itself while it cleaned its fingernails. I never do get them really clean, it said. It's a filthy job I've got but what's the use of complaining. All the same I think I'd rather have been Youth or Spring or any number of things rather than what I am. Not Youth, maybe. That's a little wet and you'd hardly get to know people before they've moved on. Spring's pretty much the same and it's a lady's job besides. Action would be nice to be, I should think.

Elsewhere Action lay in his cell smoking and looking up at the ceiling. What a career, he said. I've spent more time in the nick than anywhere else. Why couldn't I have been Death or something like that. Steady work, security.

RUSSELL HOBAN (b. 1925), *Kleinzeit*

Madam Life's a piece in bloom
  Death goes dogging everywhere:
She's the tenant of the room,
  He's the ruffian on the stair.

You shall see her as a friend,
  You shall bilk him once and twice;
But he'll trap you in the end,
  And he'll stick you for her price.

With his kneebones at your chest
  And his knuckles in your throat,
You would reason—plead—protest!
  Clutching at her petticoat;

But she's heard it all before,
  Well she knows you've had your fun,
Gingerly she gains the door,
  And your little job is done.

W. E. HENLEY (1849–1903), 'To W. R.'

Death be not proud, though some have called thee
Mighty and dreadful, for, thou art not so,
For, those, whom thou thinkst, thou dost overthrow,
Die not, poor death, nor yet canst thou kill me.
From rest and sleep, which but thy pictures be,
Much pleasure, then from thee, much more must flow,
And soonest our best men with thee do go,
Rest of their bones, and souls' delivery.
Thou art slave to Fate, Chance, kings, and desperate men,
And dost with poison, war, and sickness dwell,
And poppy, or charms can make us sleep as well,
And better than thy stroke; why swellst thou then?
One short sleep past, we wake eternally,
And death shall be no more; death, thou shalt die.

DONNE, 'Holy Sonnets'

Views and Attitudes

In a poem itself far from blank, William Empson surmises that the subject of death is one that most people should be prepared to be blank upon. Whether most people are actually blank as distinct from reticent is hard to tell. But many have been far from silent, so much is certain.

'How can we ever feel secure from death,' asks a Buddhist scripture, 'when from the womb onwards it follows us like a murderer with his sword raised to kill us?' The present section opens with a scientific-sounding rationale for fearing death: without such fear the whole species would soon come to an end. That fear itself is compellingly set out in Claudio's speech from *Measure for Measure* (any sort of life is preferable to our imaginings of death and thereafter), while the ghost of Achilles echoes the Biblical saying, 'A living dog is better than a dead lion.' We may not concur with Seneca's assertion that 'nothing that is extreme is evil', but palliatives have always been sought, of course, and found—of which the commonest is drawn from the family likeness between death and sleep.

Seneca and (more subtly) Lucretius argue that the soul dies with the body and therefore there is nothing to fear or to anticipate since there is no experiencing agent. Philip Larkin's 'Aubade' calls this

> specious stuff that says *No rational being*
> *Can fear a thing it will not feel*, not seeing
> That this is what we fear . . .

Those who lack the religious belief in an afterlife may take some consolation from the thought that those who do so believe are not invariably calm about either the act of dying or their subsequent prospects. The certain knowledge that death is universal, and not a fate reserved for you or me, suggests that we oughtn't to fuss too much about it—on the night we die a thousand others go with us, among them wise men and fools and such low forms of life as shepherds or maidservants. Anyone can do it; it can be done to anyone. Or, to frame the thought in possibly more elevating terms, death joins us to the great majority (an expression older than we may have supposed). To which considerations, however, Malraux's character rejoins: but I am not talking about death, but about *me, who is going to die*.

Since death is inevitable, how are we to dispose ourselves for it? If we are Christians, by thinking about it all the time—or so some have advised. Otherwise, by not thinking about it very much. Affairs will keep breaking in—the market price of a yoke of bullocks or, in more general terms, what Goethe's Faust, renouncing his claims on eternity and fending off diverse phantoms, called 'onward-striding'. In a wartime letter Sylvia Townsend Warner reflected on the surprise people profess at the flippant behaviour of animals in the face of death ('the live animal frisks over the dead animal and doesn't seem to know it's there'), adding: 'I don't know why they find it surprising, for in the next breath they are doing it themselves.' Yet there is truth in La Rochefoucauld's remark that death is not to be looked at steadily: our attitudes may change with circumstances, such as falling in love or having children, and especially with the passing of the years—as the changing drift of the poem mentioned on page 39 indicates with comic effect. At the same time, there can be few people, at least after a certain age, who are utterly repelled by the thought of death. Weariness of body or soul, a sense of one's work done, some 'rooted sorrow' or insoluble problem—even for the resilient and naturally cheerful these can make the prospect of an ending less than entirely unwelcome.

And ignorance is a reason both for fearing death and for not being too afraid of it.

He who pretends to look on death without fear lies. All men are afraid of dying, this is the great law of sentient beings, without which the entire human species would soon be destroyed.

JEAN-JACQUES ROUSSEAU (1712–78), *Julie, or The New Eloise*

And yet how I should prove that death is not to be feared, I cannot well tell, seeing the whole power of nature showeth that of all things death is most fearful: and to reason against nature, it were peradventure not so hard as vain. For what can reason prevail, if nature resist? It is a thing too far above man's power, to strive or to wrestle with nature, her strength passeth the might of our will, what help soever we take of reason or of authority: neither counsel nor commandment hath place, where nature doth her uttermost . . .

THOMAS LUPSET (1495–1530), *The Way of Dying Well*

I that in heill wes and gladnes,
Am trublit now with gret seiknes,
And feblit with infermitie;
Timor mortis conturbat me.

Our plesance here is all vain glory,
This fals world is but transitory,
The flesche is brukle, the Feynd is slee;
Timor mortis conturbat me.

The stait of man does change and vary,
Now sound, now seik, now blith, now sary,
Now dansand mery, now like to dee;
Timor mortis conturbat me.

No stait in Erd here standis sicker;
As with the wynd wavis the wicker,
Wavis this warldis vanitie;
Timor mortis conturbat me.

On to the dead gois all Estatis,
Princis, Prelotis, and Potestatis,
Baith rich and pur of all degree;
Timor mortis conturbat me . . .

He sparis no lord for his piscence,
Na clerk for his intelligence;
His awfull straik may no man flee;
Timor mortis conturbat me . . .

Sen he hes all my brether tane,
He will nocht lat me lif alane,
On force I mun his nixt prey be;
Timor mortis conturbat me.

Sen for the death remeid is none,
Best is that we for death dispone,
Eftir our death that lif may we;
Timor mortis conturbat me.

WILLIAM DUNBAR (*c.*1460–*c.*1520), from 'Lament for the Makers'

Before my face the picture hangs,
 That daily should put me in mind
Of those cold qualms and bitter pangs,
 That shortly I am like to find:
  But yet, alas, full little I
  Do think hereon, that I must die.

I often look upon a face
 Most ugly, grisly, bare and thin;
I often view the hollow place,
 Where eyes and nose had sometimes been;
  I see the bones across that lie,
  Yet little think that I must die.

I read the label underneath,
 That telleth me whereto I must;
I see the sentence eke that saith
 'Remember, man, that thou art dust!'
  But yet, alas, but seldom I
  Do think indeed that I must die . . .

My ancestors are turned to clay,
 And many of my mates are gone,
My youngers daily drop away,
 And can I think to 'scape alone?
  No, no, I know that I must die,
  And yet my life amend not I.

Not Solomon, for all his wit,
 Nor Samson, though he were so strong,
No king nor person ever yet
 Could 'scape, but death laid him along:
  Wherefore I know that I must die,
  And yet my life amend not I.

Though all the East did quake to hear
 Of Alexander's dreadful name,
And all the West likewise did fear
 To hear of Julius Caesar's fame,
  Yet both by death in dust now lie.
  Who then can 'scape, but he must die?

If none can 'scape death's dreadful dart,
If rich and poor his beck obey,
If strong, if wise, if all do smart,
Then I to 'scape shall have no way.
Oh! grant me grace, O God, that I
My life may mend, sith I must die.

ROBERT SOUTHWELL (1561–95), from 'Upon the Image of Death'

Ay, but to die, and go we know not where,
To lie in cold obstruction and to rot;
This sensible warm motion to become
A kneaded clod; and the delighted spirit
To bathe in fiery floods, or to reside
In thrilling region of thick-ribbed ice,
To be imprisoned in the viewless winds
And blown with restless violence round about
The pendent world; or to be worse than worst
Of those that lawless and incertain thoughts
Imagine howling: 'tis too horrible!
The weariest and most loathèd worldly life
That age, ache, penury, and imprisonment
Can lay on nature is a paradise
To what we fear of death.

WILLIAM SHAKESPEARE (1564–1616), *Measure for Measure*

Do not suppose that I do not fear death
Because I trust it is no end. You say
It must be a great comfort to live with
Such a faith, but you don't know the way
I battle on this earth

With faults of character I try to change
But they bound back on me like living things.
It is most difficult to re-arrange
Your life each day. I have known barren Springs
And death in Summer. Strange

Intimations come with Autumn. I
Would like, in certain moods, to die as leaves
Do, go back to earth. I watch the sky
Mellowing and find my strict beliefs
Hurting cruelly.

Who doesn't regret Lazarus was not
Questioned about after-lives? Of course
He only reached death's threshold. I fear what
Dark exercises may with cunning powers
Do when I am brought

To my conclusion. I have been where two
Or three died, only one of them a close
Friend. I mourn for him but he can't throw
A rope to me when that dark boatman rows
Me to what none can know.

ELIZABETH JENNINGS (b. 1926), 'The Fear of Death'

You ask me, my dear child, if I still like life. I assure you that I find many sharp pains in it; but I am still more disgusted by death: I am so unhappy at having to finish everything in this way that if I could go backwards, I would ask for nothing better. I find myself in a situation which embarrasses me: I was launched into life without my consent; I have to leave it, which I hate; and how shall I leave it?

MME DE SÉVIGNÉ, letter to her daughter, 1672

The moralist still obstinate replies,
Others' enjoyment from your woes arise,
To numerous insects shall my corpse give birth,
When once it mixes with its mother earth:
Small comfort 'tis that when Death's ruthless power
Closes my life, worms shall my flesh devour.

VOLTAIRE (1694–1778), from 'The Lisbon Earthquake', tr. Tobias Smollett, *c.*1764

I know that I shall die soon and my mind is reconciled to it; but when I think that my body will be put into a coffin, that the lid of the coffin will be screwed down and I will be buried under earth, I am horrified. I am well aware that my horror is unreasonable, that I shall not be feeling anything by then, but I cannot overcome this feeling. Sometimes I also have the feeling—and that is also unreasonable—that I shall not die. I read somewhere that a Frenchman had begun his will with the following words: 'Not *when* but *if* I should die one day . . .'

ALEXANDER SERGEYEVICH BUTURLIN, 1915; quoted by Sergei Tolstoy, *Tolstoy Remembered by his Son*, tr. Moura Budberg

Do you ever think of yourself as actually *dead*, lying in a box with a lid on it? . . . It's silly to be depressed by it. I mean one thinks of it like being *alive* in a box, one keeps forgetting to take into account the fact that one is *dead* . . . which should make a difference . . . shouldn't it? I mean, you'd never *know* you were in a box, would you? It would be just like being *asleep* in a box. Not that I'd like to sleep in a box, mind you, not without any air—you'd wake up dead, for a start, and then where would you be? Apart from inside a box . . . Because you'd be helpless, wouldn't you? Stuffed in a box like that, I mean you'd be in there for ever. Even taking into account the fact that you're dead, really . . . I wouldn't think about it, if I were you. You'd only get depressed. (*Pause.*) Eternity is a terrible thought. I mean, where's it going to end?

TOM STOPPARD (b. 1937), *Rosencrantz and Guildenstern Are Dead*

'Dying,' he said to me, 'is a very dull, dreary affair.' Suddenly he smiled. 'And my advice to you is to have nothing whatever to do with it,' he added.

SOMERSET MAUGHAM, shortly before his death in 1965, as recorded by his nephew, Robin Maugham

After death nothing is, and nothing, death:
The utmost limit of a gasp of breath.
Let the ambitious zealot lay aside
His hopes of heaven, whose faith is but his pride;
  Let slavish souls lay by their fear,
  Nor be concerned which way nor where
  After this life they shall be hurled.
Dead, we become the lumber of the world,
And to that mass of matter shall be swept
Where things destroyed with things unborn are kept.
  Devouring time swallows us whole;
Impartial death confounds body and soul.
  For Hell and the foul fiend that rules
  God's everlasting fiery jails
  (Devised by rogues, dreaded by fools),
With his grim, grisly dog that keeps the door,
  Are senseless stories, idle tales,
    Dreams, whimseys, and no more.

SENECA, *Troades*, tr. John Wilmot, Earl of Rochester, 1680

What has this Bugbear Death to frighten Man,
If Souls can die, as well as Bodies can?
For, as before our Birth we felt no Pain,
When Punic arms infested Land and Main,
When Heaven and Earth were in confusion hurl'd,
For the debated Empire of the World,
Which aw'd with dreadful expectation lay,
Sure to be Slaves, uncertain who should sway:
So, when our mortal frame shall be disjoin'd,
The lifeless Lump uncoupled from the mind,
From sense of grief and pain we shall be free;
We shall not feel, because we shall not *Be* . . .
Nay, ev'n suppose when we have suffer'd Fate,
The Soul could feel, in her divided state,
What's that to us? for we are only we
While Souls and Bodies in one frame agree . . .
We, who are dead and gone, shall bear no part
In all the pleasures, nor shall feel the smart,
Which to that other Mortal shall accrue,
Whom, of our Matter Time shall mould anew.
For backward if you look, on that long space
Of Ages past, and view the changing face
Of Matter, toss'd and variously combin'd
In sundry shapes, 'tis easy for the mind
From thence t'infer, that Seeds of things have been
In the same order as they now are seen:
Which yet our dark remembrance cannot trace,
Because a pause of Life, a gaping space,
Has come betwixt, where memory lies dead,
And all the wand'ring motions from the sense are fled.
For whosoe'er shall in misfortunes live,
Must *Be*, when those misfortunes shall arrive;
And since the Man who *Is* not, feels not woe
(For death exempts him and wards off the blow,
Which we, the living, only feel and bear),
What is there left for us in Death to fear?
When once that pause of life has come between,
'Tis just the same as we had never been.

LUCRETIUS (?94–55 BC), *De Rerum Natura*, tr. John Dryden

Men fear Death as children fear to go in the dark; and as that natural fear in children is increased with tales, so is the other.

A man would die, though he were neither valiant nor miserable, only upon a weariness to do the same thing so oft over and over.

FRANCIS BACON (1561–1626), *Of Death*

Where dwells that wish most ardent of the wise?
Too dark the sun to see it; highest stars
Too low to reach it; death, great death alone,
O'er stars and sun, triumphant, lands us there.
  Nor dreadful our transition; tho' the mind,
An artist at creating self-alarms,
Rich in expedients for inquietude,
Is prone to paint it dreadful. Who can take
Death's portrait true? The tyrant never sat.
Our sketch all random strokes, conjecture all;
Close shuts the grave, nor tells one single tale.
Death, and his image rising in the brain,
Bear faint resemblance; never are alike;
Fear shakes the pencil; fancy loves excess;
Dark ignorance is lavish of her shades:
And these the formidable picture draw.

EDWARD YOUNG (1683–1765), *Night Thoughts on Life, Death, and Immortality*

If it be a short and violent death, we have no leisure to fear it; if otherwise, I perceive that according as I engage myself in sickness, I do naturally fall into some disdain and contempt of life.

It is uncertain where death looks for us; let us expect her everywhere: the premeditation of death, is a forethinking of liberty. He who hath learned to die, hath unlearned to serve. There is no evil in life, for him that hath well conceived, how the privation of life is no evil. To know how to die, doth free us from all subjection and constraint.

MICHEL DE MONTAIGNE (1533–92), *Essays*, tr. John Florio, 1603

It is impossible that anything so natural, so necessary, and so universal as death, should ever have been designed by Providence as an evil to mankind.

JONATHAN SWIFT (1667–1745), 'Thoughts on Religion'

Mortal man, you have been a citizen in this great City; what does it matter to you whether for five or fifty years? For what is according to its law is equal for every man. Why is it hard, then, if Nature who brought you in, and no despot nor unjust judge, sends you out of the City—as though the master of the show, who engaged an actor, were to dismiss him from the stage? 'But I have not spoken my five acts, only three.' 'What you say is true, but in life three acts are the whole play.' For He determines the perfect whole, the cause yesterday of your composition, today of your dissolution; you are the cause of neither. Leave the stage, therefore, and be reconciled, for He also who lets his servant depart is reconciled.

MARCUS AURELIUS (121–80), *Meditations*, tr. A. S. L. Farquharson

Whoever has lived long enough to find out what life is, knows how deep a debt of gratitude we owe to Adam, the first great benefactor of our race. He brought death into the world.

All say, 'How hard it is that we have to die'—a strange complaint to come from the mouths of people who have had to live.

MARK TWAIN (1835–1910), *Pudd'nhead Wilson*

O death, how bitter is the remembrance of thee to a man that is at peace in his possessions, unto the man that hath nothing to distract him, and hath prosperity in all things, and that still hath strength to receive meat! O death, acceptable is thy sentence unto a man that is needy, and that faileth in strength, that is in extreme old age, and is distracted about all things, and is perverse, and hath lost patience! Fear not the sentence of death; remember them that have been before thee, and that come after: this is the sentence from the Lord over all flesh. And why dost thou refuse, when it is the good pleasure of the Most High? Whether it be ten, or a hundred, or a thousand years, there is no inquisition of life in the grave.

Ecclesiasticus, 41

Death is not an evil, for it liberates from all evils, and if it deprives man of any good thing it also takes away his desire for it. Old age is the supreme evil, for it deprives man of all pleasures, while leaving him appetites for them; and brings with it all sufferings. Nevertheless, men fear death and desire old age.

GIACOMO LEOPARDI (1798–1837), *Operette morali*, tr. James Thomson, 'B.V.', 1867

It is a thing that every one suffers, even persons of the lowest resolution, of the meanest virtue, of no breeding, of no discourse. Take away but the pomps of death, the disguises and solemn bugbears, the tinsel, and the actings by candle-light, and proper and fantastic ceremonies, the minstrels and the noise-makers, the women and the weepers, the swoonings and the shriekings, the nurses and the physicians, the dark room and the ministers, the kindred and the watchers; and then to die is easy, ready and quitted from its troublesome circumstances. It is the same harmless thing that a poor shepherd suffered yesterday, or a maidservant today; and at the same time in which you die, in that very night a thousand creatures die with you, some wise men, and many fools; and the wisdom of the first will not quit him, and the folly of the latter does not make him unable to die.

Of all the evils of the world which are reproached with an evil character, death is the most innocent of its accusation. For when it is present, it hurts nobody; and when it is absent, it is indeed troublesome, but the trouble is owing to our fears, not the affrighting and mistaken object: and besides this, if it were an evil, it is so transient, that it passes like the instant or undiscerned portion of the present time; and either it is past, or it is not yet; for just when it is, no man hath reason to complain of so insensible, so sudden, so undiscerned a change.

JEREMY TAYLOR (1613–67), *Holy Dying*

One should part from life as Ulysses parted from Nausicaa—blessing it rather than in love with it.

FRIEDRICH NIETZSCHE (1844–1900), *Beyond Good and Evil*, tr. Helen Zimmern

Perhaps the best cure for the fear of death is to reflect that life has a beginning as well as an end. There was a time when we were not: this gives us no concern—why then should it trouble us that a time will come when we shall cease to be? I have no wish to have been alive a hundred years ago, or in the reign of Queen Anne: why should I regret and lay it so much to heart that I shall not be alive a hundred years hence, in the reign of I cannot tell whom? . . . To die is only to be as we were before we were born; yet no one feels any remorse, or regret, or repugnance, in contemplating this last idea. It is rather a relief and disburthening of the mind: it seems to have been holiday-time with us then: we were not called to appear upon the stage of life, to wear robes or tatters, to laugh or cry, be hooted or applauded; we had lain *perdus* all this while, snug, out of harm's way; and had slept out our thousands of centuries without wanting to be waked up; at peace and free from care, in a long nonage, in a sleep deeper and calmer than that of infancy, wrapped in the finest and softest dust. And the worst that we dread is, after a short, fretful, feverish being, after vain hopes, and idle fears, to sink to final repose again, and forget the troubled dream of life!

WILLIAM HAZLITT (1778–1830), 'On the Fear of Death', *Table Talk*

I am quite ready to admit that I ought to be grieved at death, if I were not persuaded in the first place that I am going to other gods who are wise and good (of which I am as certain as I can be of any such matters), and secondly (though I am not so sure of this last) to men departed, better than those whom I leave behind; and therefore I do not grieve as I might have done, for I have good hope that there is yet something remaining for the dead, and as has been said of old, some far better thing for the good than for the evil.

Many a man has been willing to go to the world below animated by the hope of seeing there an earthly love, or wife, or son, and conversing with them. And will he who is a true lover of wisdom, and is strongly persuaded in like manner that only in the world below he can worthily enjoy her, still repine at death? Will he not depart with joy? Surely he will, O my friend, if he be a true philosopher. For he will have a firm conviction that there, and there only, he can find wisdom in her purity.

PLATO (?429–347 BC), *Phaedo*, tr. Benjamin Jowett

Life is the desert, life the solitude;
Death joins us to the great majority:
'Tis to be born to Plato's and to Caesar;
'Tis to be great for ever;
'Tis pleasure, 'tis ambition, then, to die.

EDWARD YOUNG, *The Revenge*

This olde man gan looke in his visage,
And seyde thus, 'For I ne can nat finde
A man, though that I walkèd into Inde,
Neither in citee ne in no village,

That wolde chaunge his youthe for myn age;
And therfore moot I han myn age stille,
As longe time as it is Goddes wille.
Ne Deeth, allas! ne wol nat han my lyf;
Thus walke I, lyk a restelees caityf,
And on the ground, which is my modres gate,
I knokke with my staf, bothe erly and late,
And seye, "Leve moder, leet me in!
Lo, how I vanish, flesh, and blood, and skin!
Allas! whan shul my bonès been at reste?
Moder, with yow wolde I chaunge my cheste,
That in my chambre longe tyme hath be,
Ye, for an heyre clout to wrappe me!"
But yet to me she wol nat do that grace,
For which ful pale and welkèd is my face.'

GEOFFREY CHAUCER (?1340–1400), from 'The Pardoner's Tale'

I know this body but a sink of folly,
The ground-work and rais'd frame of woe and frailty,
The bond and bundle of corruption,
A quick corse, only sensible of grief,
A walking sepulchre, or household thief,
A glass of air, broken with less than breath,
A slave bound face to face to Death till death . . .
I know, besides,
That life is but a dark and stormy night
Of senseless dreams, terrors, and broken sleeps;
A tyranny, devising pains to plague
And make man long in dying, racks his death;
And Death is nothing; what can you say more?
I being a large globe, and a little earth,
Am seated like earth, betwixt both the heavens,
That if I rise, to heaven I rise; if fall,
I likewise fall to heaven; what stronger faith
Hath any of your souls? What say you more?

GEORGE CHAPMAN (1559–1634), *The Tragedy of Charles Duke of Byron*

There is a land of pure delight,
Where saints immortal reign;
Infinite day excludes the night,
And pleasures banish pain.

There everlasting spring abides,
  And never-withering flowers:
Death, like a narrow sea, divides
  This heavenly land from ours.

Sweet fields beyond the swelling flood
  Stand dress'd in living green:
So to the Jews old Canaan stood,
  While Jordan roll'd between.

But timorous mortals start and shrink
  To cross the narrow sea,
And linger shivering on the brink,
  And fear to launch away.

O could we make our doubts remove,
  These gloomy doubts that rise,
And see the Canaan that we love
  With unbeclouded eyes,

Could we but climb where Moses stood,
  And view the landscape o'er,
Not Jordan's stream, nor Death's cold flood,
  Should fright us from the shore.

ISAAC WATTS (1674–1748), 'A Prospect of Heaven'

Clother of the lily, Feeder of the sparrow,
  Father of the fatherless, dear Lord,
Tho' Thou set me as a mark against Thine arrow,
  As a prey unto Thy sword,
As a plough'd-up field beneath Thy harrow,
  As a captive in Thy cord,
Let that cord be love; and some day make my narrow
  Hallow'd bed according to Thy Word. Amen.

CHRISTINA ROSSETTI (1830–94), 'A Prayer'

I know the truth—give up all other truths!
No need for people anywhere on earth to struggle.
Look—it is evening, look, it is nearly night:
what do you speak of, poets, lovers, generals?

The wind is level now, the earth is wet with dew,
the storm of stars in the sky will turn to quiet.
And soon all of us will sleep under the earth, we
who never let each other sleep above it.

MARINA TSVETAYEVA (1892–1941), 'I know the truth', tr. Elaine Feinstein, with Angela Livingstone

Over every hill
All is still;
In no leaf of any tree
Can you see
The motion of a breath;
Every bird has ceased its song.
Wait; and thou too, ere long,
Shall be quiet in death.

GOETHE, 'Wanderer's Night-Song', tr. Arthur Hugh Clough

For to him that is joined to all the living there is hope: for a living dog is better than a dead lion. For the living know that they shall die: but the dead know not any thing, neither have they any more a reward; for the memory of them is forgotten. Also their love, and their hatred, and their envy, is now perished; neither have they any more a portion for ever in any thing that is done under the sun.

Ecclesiastes, 9

Life is, as I've said since I was 10, awfully interesting—if anything, quicker, keener at 44 than 24—more desperate I suppose, as the river shoots to Niagara—my new vision of death; active, positive, like all the rest, exciting; & of great importance—as an experience.

'The one experience I shall never describe' I said to Vita yesterday.

VIRGINIA WOOLF (1882–1941), *Diary*, 1926

Then there is this civilizing love of death, by which
Even music and painting tell you what else to love.
Buddhists and Christians contrive to agree about death

Making death their ideal basis for different ideals.
The Communists however disapprove of death
Except when practical. The people who dig up

Corpses and rape them are I understand not reported.
The Freudians regard the death-wish as fundamental,
Though 'the clamour of life' proceeds from its rival 'Eros'.

Whether you are to admire a given case for making less clamour
Is not their story. Liberal hopefulness
Regards death as a mere border to an improving picture.

Because we have neither hereditary nor direct knowledge of death
It is the trigger of the literary man's biggest gun
And we are happy to equate it to any conceived calm.

Heaven me, when a man is ready to die about something
Other than himself, and is in fact ready because of that,
Not because of himself, that is something clear about himself.

Otherwise I feel very blank upon this topic,
And think that though important, and proper for anyone to bring up,
It is one that most people should be prepared to be blank upon.

WILLIAM EMPSON (b. 1906), 'Ignorance of Death'

Certain, 'tis certain; very sure, very sure: death, as the Psalmist saith, is certain to all; all shall die. How a good yoke of bullocks at Stamford fair?

SHAKESPEARE, *Henry IV, Part Two*

Beneath this stone, in hope of Zion,
Doth lie the landlord of the 'Lion'.
His son keeps on the business still,
Resign'd unto the Heavenly will.

In the churchyard of Upton-on-Severn

He spake, to whom I, answ'ring, thus replied:
'O Peleus' son! Achilles! bravest far
Of all Achaia's race! I here arrived
Seeking Tiresias, from his lips to learn,
Perchance, how I might safe regain the coast
Of craggy Ithaca; for tempest-toss'd

Perpetual, I have neither yet approach'd
Achaia's shore, or landed on my own.
But as for thee, Achilles! never man
Hath known felicity like thine, or shall,
Whom living we all honour'd as a God,
And who maintain'st, here resident, supreme
Control among the dead; indulge not then,
Achilles, causeless grief that thou hast died.'
I ceased, and answer thus instant received:
'Renown'd Ulysses! think not death a theme
Of consolation; I had rather live
The servile hind for hire, and eat the bread
Of some man scantily himself sustain'd,
Than sov'reign empire hold o'er the shades.'

HOMER (?8th century BC), *Odyssey*, tr. William Cowper

A poor old Woodman with a leafy load,
Both with his faggots and his years bowed down,
Groaning, with heavy steps along the road
Laboured to reach his hovel smoked and brown.
At last, with toil and pain exhausted quite,
He dropt the boughs to muse upon his plight.
What pleasure had he known since he was born?
In th' whole round world was any more forlorn?
Often no bread, never an hour of rest:
His wife, his children, soldiers foraging,
Forced labour, debt, the taxes for the King—
The finished picture of a life unblest!
He called on Death, who instantly stood by,
And asked of him: 'What is 't you lack?'
'I wanted you,' was the reply,
'To help me load this wood upon my back.'

Death can take all ills away;
But here we are, and here would stay.
Better to suffer than to die
Is Everyman's philosophy.

JEAN DE LA FONTAINE (1621–95), 'Death and the Woodman', tr. Edward Marsh

The hop-poles stand in cones,
  The icy pond lurks under,
The pole-tops steeple to the thrones
  Of stars, sound gulfs of wonder;
But not the tallest there, 'tis said,
Could fathom to this pond's black bed.

Then is not death at watch
  Within those secret waters?
What wants he but to catch
  Earth's heedless sons and daughters?
With but a crystal parapet
Between, he has his engines set.

Then on, blood shouts, on, on,
  Twirl, wheel and whip above him,
Dance on this ball-floor thin and wan,
  Use him as though you love him;
Court him, elude him, reel and pass,
And let him hate you through the glass.

EDMUND BLUNDEN (1896–1974), 'The Midnight Skaters'

Do not go gentle into that good night,
Old age should burn and rave at close of day;
Rage, rage against the dying of the light.

Though wise men at their end know dark is right,
Because their words had forked no lightning they
Do not go gentle into that good night.

Good men, the last wave by, crying how bright
Their frail deeds might have danced in a green bay,
Rage, rage against the dying of the light.

Wild men who caught and sang the sun in flight,
And learn, too late, they grieved it on its way,
Do not go gentle into that good night.

Grave men, near death, who see with blinding sight
Blind eyes could blaze like meteors and be gay,
Rage, rage against the dying of the light.

And you, my father, there on the sad height,
Curse, bless, me now with your fierce tears, I pray.
Do not go gentle into that good night,
Rage, rage against the dying of the light.

DYLAN THOMAS (1914–53), 'Do not go gentle into that good night'

If Rabbi Hillel was right in saying that death presents various aspects only to the unwise, then indeed very few human beings are really wise, for to most persons the aspect varies from time to time, according to moods and circumstances. Age certainly often modifies the aspect, as Mr Edmond G. A. Holmes recently stated that it had done in his own case. It obliged him to alter many of the stanzas in his poem, 'To Death', included in *The Creed of My Heart and Other Poems*, London, 1912.

F. PARKES WEBER, *Aspects of Death in Art and Epigram*, 1914

There is . . . no death . . . There is only . . . *me* . . . *me* . . . *who is going to die* . . .

ANDRÉ MALRAUX (1901–76), *The Royal Way*

I ask myself, why am I so frightened of an atomic war wiping out the bulk of mankind? It is true that I am frightened, and deeply upset, by the thought—so deeply that I have to ration myself how long I think about it, no more than a few seconds at a time. And I guess that most people are the same. Yet, after all, whereas an atomic war would leave *some* humans alive, my death, so far as I am concerned, would leave none whatever . . . It can only be, I think, that I would feel even more intensely about my own coming death, were it not that I am prevented from picturing it.

P. N. FURBANK (b. 1920), 'A Note on Death'

Death and the sun are not to be looked at steadily.

LA ROCHEFOUCAULD (1613–80), *Maxims*

'Life's too short for worrying.'
'Yes, that's what worries me.'

Origin unknown

I'm not afraid to die. I just don't want to be there when it happens.

WOODY ALLEN (b. 1935)

Lord, let me know mine end, and the number of my days: that I may be certified how long I have to live (Ps. xxxix, 4).

Of all prayers this is the insanest. That the one who uttered it should have made and retained a reputation is a strong argument in favour of his having been surrounded with courtiers. 'Lord, let me not know mine end' would be better, only it would be praying for what God has already granted us. 'Lord, let me know A.B.'s end' would be bad enough. Even though A.B. were Mr Gladstone—we might hear he was not to die yet. 'Lord, stop A.B. from knowing my end' would be reasonable, if there were any use in praying that A.B. might not be able to do what he never can do. Or can the prayer refer to the other end of life and mean 'Lord, let me know my beginning'? This again would not be always prudent.

The prayer is a silly piece of petulance and it would have served the maker of it right to have had it granted. 'Cancer in about three months after great suffering' or 'Ninety, a burden to yourself and every one else'—there is not so much to pick and choose between them. Surely, 'I thank thee, O Lord, that thou hast hidden mine end from me' would be better.

SAMUEL BUTLER (1835–1902), *Notebooks*

Man is only a reed, the weakest thing in nature—but a thinking reed. It does not take the universe in arms to crush him; a vapour, a drop of water, is enough to kill him. But, though the universe should crush him, man would still be nobler than his destroyer, because he knows that he is dying, that the universe has the advantage of him; the universe knows nothing of this.

BLAISE PASCAL (1623–62), *Pensées*

But for your Terror
Where would be Valour?
What is Love for
  But to stand in your way?
Taker and Giver,
For all your endeavour
You leave us with more
  Than you touch with decay.

OLIVER ST JOHN GOGARTY (1878–1957), 'To Death'

I was of delicate mind. I stepped aside for my needs,
Disdaining the common office. I was seen from afar and killed . . .
How is this matter for mirth? Let each man be judged by his deeds.
I have paid my price to live with myself on the terms that I willed.

RUDYARD KIPLING (1865–1936), 'The Refined Man'

The dictator-hero can grind down his citizens till they are all alike, but he cannot melt them into a single man. This is beyond his power. He can order them to merge, he can incite them to mass-antics, but they are obliged to be born separately, and to die separately, and, owing to these unavoidable termini, will always be running off the totalitarian rails. The memory of birth and the expectation of death always lurk within the human being, making him separate from his fellows and consequently capable of intercourse with them.

E. M. FORSTER (1879–1970), 'What I Believe'

One summer only grant me, you powerful Fates,
And one more autumn only for mellow song,
So that more willingly, replete with
Music's late sweetness, my heart may die then.

The soul in life denied its god-given right
Down there in Orcus also will find no peace;
But when what's holy, dear to me, the
Poem's accomplished, my art perfected,

Then welcome, silence, welcome cold world of shades!
I'll be content, though here I must leave my lyre
And songless travel down; for *once* I
Lived like the gods, and no more is needed.

FRIEDRICH HÖLDERLIN (1770–1843), 'To the Fates', tr. Michael Hamburger

It may be only glory that we seek here, but I persuade myself that, as long as we remain here, that is right. Another glory awaits us in heaven and he who reaches there will not wish even to think of earthly fame. So this is the natural order, that among mortals the care of things mortal should come first; to the transitory will then succeed the eternal; from the first to the second is the natural progression.

FRANCESCO PETRARCH (1304–74), *Secretum*, tr. W. H. Draper

Cowards die many times before their deaths;
The valiant never taste of death but once.
Of all the wonders that I yet have heard,
It seems to me most strange that men should fear,
Seeing that death, a necessary end,
Will come when it will come.

SHAKESPEARE, *Julius Caesar*

The Hour of Death

THE view expressed by Erasmus's Phaedrus that the act of dying is meant to seem dreadful in order to discourage us from suicide is in line with Rousseau's theory cited in the preceding section. Aversion to death is necessary to the survival of the species, rather in the spirit of the French proverb to the effect that fear of the gendarme is the beginning of wisdom. In case Erasmus seems to exaggerate the likely incidence of suicide in the absence of such restraints, let me adduce the schoolboy in Japan who threw himself in front of a train because he had been reprimanded by the teacher in class. Loss of face can seem a more serious thing than loss of life.

The passage from Jung confirms the remarks in the preface to the foregoing section on age as mitigating our fear of death. E. M. Forster's depiction of Mrs Moore in *A Passage to India* illustrates one form of gradual withdrawal from life, a less exalted one than Yeats's but surely commoner—an impatience with the affairs of others, a weary wish to be left alone, to 'finish my duties and be gone'. 'She had come to that state where the horror of the universe and its smallness are both visible at the same time—the twilight of the double vision in which so many elderly people are involved.' Jung says, 'Nature herself is already preparing for the end', though we may be less convinced by his claim that the mentally healthy or un-neurotic elderly will focus on the goal of death. The word 'goal' suggests something altogether more deliberate and energetic than a more or less quiescent disengagement from aims, aspirations and preoccupations.

Leopardi's embalmed informant offers a comforting account of the severing of the links with life: a dying person feels but little, and indeed, simply through being weak, his feelings may resemble those of pleasure! As would be expected, Lawrence and Whitman see the process of departure as a more intricate and strenuous matter, while a Zen master has dismissed the whole affair as an illusion: one does not come, therefore one cannot go.

Were he a maker of books, Montaigne stated modestly, he would compile a record of diverse deaths. Death-bed scenes are abundant in literature, and only a sparse but (I hope) diverse selection is presented here, ranging from the urbane farewell of Socrates, the gallant bearing of the Duchess of Malfi, the passing of Lear (compared by those present to the end of the world), to the disreputable exit of old Thomas Parr and the gracelessly business-like finis of M. Grandet, the supererogatory death of Dr Marjoribanks, the lucid last hours reported by Rilke, the last agonies imagined by John Betjeman, and the more cheerful prospects held out by some other writers. 'I lie back on my pillows and sleep with my face to the South.' Despite Johnson's dismissing how a man dies as of no importance, most of us have no wish to be reprimanded on our death-beds.

~

There is nothing I desire more to be informed of, than of the death of men: that is to say, what words, what countenance, and what face they show at their death . . . Were I a composer of books, I would keep a register, commented of the diverse deaths, which in teaching men to die, should after teach them to live.

MONTAIGNE, *Essays*

Remember now thy Creator in the days of thy youth, while the evil days come not, nor the years draw nigh, when thou shalt say, I have no pleasure in them; While the sun, or the light, or the moon, or the stars, be not darkened, nor the clouds return after the rain: In the day when the keepers of the house shall tremble, and the strong men shall bow themselves, and the grinders cease because they are few, and those that look out of the windows be darkened, And the doors shall be shut in the streets, when the sound of the grinding is low, and he shall rise up at the voice of the bird, and all the daughters of music shall be brought low; Also when they shall be afraid of that which is high, and fears shall be in the way, and the almond tree shall flourish, and the grasshopper shall be a burden, and desire shall fail: because man goeth to his long home, and the mourners go about the streets: Or ever the silver cord be loosed, or the golden bowl be broken, or the pitcher be broken at the fountain, or the wheel broken at the cistern. Then shall the dust return to the earth as it was: and the spirit shall return unto God who gave it.

Ecclesiastes, 12

Death does not leave us body enough to occupy any space, and only the tombs there preserve some shape; our flesh soon changes its nature, our body takes another name; even that of *corpse*, says Tertullian, for this still suggests some sort of human form, does not stay with us long; it becomes a *je ne sais quoi* which no longer has a name in any language; so true is it that everything dies in it, even to those funereal terms by which its miserable remains were known!

BOSSUET, *Oraison funèbre d'Henriette d'Angleterre, Duchesse d'Orléans*, 1670

Life is an energy-process. Like every energy-process, it is in principle irreversible and is therefore directed towards a goal. That goal is a state of rest. In the long run everything that happens is, as it were, no more than the initial disturbance of a perpetual state of rest which forever attempts to re-establish itself . . . Thoughts of death pile up to an astonishing degree as the years increase. Willy-nilly, the ageing person prepares himself for death. That is why I think that nature herself is already preparing for the end. Objectively it is a matter of indifference what the individual consciousness may think about it. But subjectively it makes an enormous difference whether consciousness keeps in step with the psyche or whether it clings to opinions of which the heart knows nothing. It is just as neurotic in old age not to focus upon the goal of death as it is in youth to repress fantasies which have to do with the future.

C. G. JUNG (1875–1961), 'The Soul and Death', *The Structure and Dynamics of the Psyche*, tr. R. F. C. Hull

Now shall I make my soul,
Compelling it to study
In a learned school
Till the wreck of body,
Slow decay of blood,
Testy delirium
Or dull decrepitude,
Or what worse evil come—
The death of friends, or death
Of every brilliant eye
That made a catch in the breath—
Seem but the clouds of the sky
When the horizon fades;
Or a bird's sleepy cry
Among the deepening shades.

W. B. YEATS (1865–1939), from 'The Tower'

O Lord, grant each his own, his death indeed,
the dying which out of that same life evolves
in which he once had meaning, love, and need.

RAINER MARIA RILKE (1875–1926), *The Book of Hours*, tr. J. B. Leishman

Marcolphus. I've never had the experience of being present at a deathbed.

Phaedrus. I have—oftener than I would wish.

Marc. But is death as horrible a thing as it's commonly asserted to be?

Phaed. The road leading up to it is harder than death itself. If a man dismisses from his thought the horror and imagination of death, he will have rid himself of a great part of the evil. In brief, whatever the torment of sickness or death, it is rendered much more endurable if a person surrenders himself wholly to the divine will. For awareness of death, when the soul is already separated from the body, is, I think, either non-existent or else an extremely low-grade awareness, because before Nature reaches this point it dulls and stuns all areas of sensation.

Marc. We're born without our being aware of it.

Phaed. But not without a mother's being aware of it.

Marc. Why don't we die in the same way? Why did God mean death to be such a torture?

Phaed. He meant birth to be painful and dangerous for the mother in order that she might love her offspring more dearly. But death he meant to be dreadful for everyone, lest men far and wide commit suicide. And since, even today, we see so many do violence to themselves, what do you suppose would happen if death weren't horrible? Whenever a servant or even a young son got a thrashing, whenever a wife fell out with her husband, whenever a man lost his money, or something else occurred that upset him, off they'd rush to noose, sword, river, cliff, poison. As matters now stand, the bitterness of death makes life more precious to us, especially since physicians can't cure a man once he's dead. Though all men don't die in the same way, just as they're not all born in the same way. Some find release in a quick death; others sink into a slow death. Persons with sleeping sickness, likewise those stung by an adder, die unconscious, stupefied with sleep. I've noticed this: that no kind of death is so bitter but that it can be endured if one has resolved to die with steadfast mind.

DESIDERIUS ERASMUS (?1466–1536), *Colloquies*, tr. Craig R. Thompson

Ruysch. A thousand questions to put to you come into my mind. But as the time is short, and leaves no space for choice, let me know briefly how you felt in body and mind at the moment of death.

Mummy. I was not aware of the exact moment of death.

The other Mummies. Nor were we.

R. How not aware?

M. For example, as you are never aware of the moment you begin to sleep, however much attention you pay.

R. But falling asleep is natural.

M. And does not dying appear to you natural? Show me a man, a beast, or a plant, that dies not.

R. I no longer wonder that you go on singing and speaking, if you were not aware of death.

Thus he, unaware of the blow, it is said,
Went on fighting, and was dead;

writes an Italian poet. I thought that as to this business of death those like you would know something more than the living. But then, to be serious once more, did you feel no pain in the instant of death?

M. What pain can that be of which he who suffers it is not aware?

R. At any rate, all are convinced that the feeling of death is most painful.

M. As if death were a feeling, and not rather the contrary . . . if man cannot be conscious of the moment in which the vital operations, in a greater or less degree, are only interrupted, be it through sleep or lethargy or syncope or whatever cause; how shall he be conscious of that in which the same operations altogether cease, and not for a short space of time, but for ever? And besides, how can it be that a vivid feeling should exist in death? much more, that death itself should be by its very nature a vivid feeling? When the sentient faculty is not only debilitated and small, but reduced so low that it fails and perishes, do you believe that the person is capable of a strong feeling? much more, do you believe that this very dying out of the faculty of feeling must be a very great feeling? . . .

R. Then what is death, if it is not anguish?

M. Rather pleasure than otherwise. Know that dying, like falling asleep, is not instantaneous but gradual. It is true that the degrees are more or fewer, greater or less, according to the variety of the causes and kinds of death. In the last moment death brings no pain or pleasure whatever, any more than sleep. In the preceding moments it cannot produce pain: for pain is vivid; and the feelings of man in that hour, that is when death has commenced, are moribund, which is as much as to say extremely reduced in force. It may well be cause of pleasure: for pleasure is not always vivid; indeed, perhaps the greater part of human delights consists in some sort of languor . . . For myself, although indeed in the hour of death I did not pay much attention to what I felt, because the doctors ordered me not to fatigue my brain; I nevertheless remember that the feeling I experienced was not very different from the satisfaction produced in men by the languor of sleep, during the time they are falling asleep.

LEOPARDI, 'Dialogue between Frederic Ruysch and his Mummies', tr. James Thomson

Ah! gentle, fleeting, wav'ring sprite,
Friend and associate of this clay!
  To what unknown region borne,
Wilt thou now wing thy distant flight?
No more with wonted humour gay,
  But pallid, cheerless, and forlorn.

EMPEROR HADRIAN (76–138), 'Hadrian's Address to his Soul when Dying', tr. Byron

At the last, tenderly,
From the walls of the powerful fortress'd house,
From the clasp of the knitted locks, from the keep of the well-closed doors,
Let me be wafted.

Let me glide noiselessly forth;
With the key of softness unlock the locks—with a whisper,
Set ope the doors O soul.

Tenderly—be not impatient,
(Strong is your hold O mortal flesh,
Strong is your hold O love.)

WALT WHITMAN (1819–92), 'The Last Invocation'

Be careful, then, and be gentle about death.
For it is hard to die, it is difficult to go through
the door, even when it opens.

And the poor dead, when they have left the walled
and silvery city of the now hopeless body
where are they to go, Oh where are they to go?

They linger in the shadow of the earth.
The earth's long conical shadow is full of souls
that cannot find the way across the sea of change.

Be kind, Oh be kind to your dead
and give them a little encouragement
and help them to build their little ship of death.

For the soul has a long, long journey after death
to the sweet home of pure oblivion.
Each needs a little ship, a little ship
and the proper store of meal for the longest journey.

Oh, from out of your heart
provide for your dead once more, equip them
like departing mariners, lovingly.

D. H. Lawrence (1885–1930), 'All Souls' Day'

Betray no surprise,
Let the eyelids slip down
Until they become
Made of true stone.

Leave it all to the heart,
Although it should stop.
It beats for itself alone
On its secret slope.

The hands will stretch out
In their boat of ice,
And the forehead be bare
—Between armies, and void,
Like a great public square.

Jules Supervielle (1884–1966), 'Whisper in agony' (original title)

I was not aware of the moment when I first crossed the threshold of this life.

What was the power that made me open out into this vast mystery like a bud in the forest at midnight!

When in the morning I looked upon the light I felt in a moment that I was no stranger in this world, that the inscrutable without name and form had taken me in its arms in the form of my own mother.

Even so, in death the same unknown will appear as ever known to me. And because I love this life, I know I shall love death as well.

The child cries out when from the right breast the mother takes it away, in the very next moment to find in the left one its consolation.

RABINDRANATH TAGORE (1861–1941), *Gitanjali*

The Soul's dark cottage, batter'd and decay'd,
Lets in new light through chinks that time has made;
Stronger by weakness, wiser men become,
As they draw near to their eternal home:
Leaving the Old, both worlds at once they view,
That stand upon the threshold of the New.

EDMUND WALLER (1606–87), from 'Of the Last Verses in the Book'

Because I could not stop for Death—
He kindly stopped for me—
The Carriage held but just Ourselves—
And Immortality.

We slowly drove—He knew no haste
And I had put away
My labour and my leisure too,
For His Civility—

We passed the School, where Children strove
At Recess—in the Ring—
We passed the Fields of Gazing Grain—
We passed the Setting Sun—

Or rather—He passed Us—
The Dews drew quivering and chill—
For only Gossamer, my Gown—
My Tippet—only Tulle—

We paused before a House that seemed
A Swelling of the Ground—
The Roof was scarcely visible—
The Cornice—in the Ground—

Since then—'tis Centuries—and yet
Feels shorter than the Day
I first surmised the Horses' Heads
Were toward Eternity—

EMILY DICKINSON (1830–86), 'Because I could not stop for Death'

With courage seek the kingdom of the dead;
The path before you lies,
It is not hard to find, nor tread;
No rocks to climb, no lanes to thread;
But broad, and straight, and even still,
And ever gently slopes downhill;
You cannot miss it, though you shut your eyes.

LEONIDAS OF TARENTUM (*fl.* *c.*274 BC), tr. Charles Merivale

Does the road wind uphill all the way?
Yes, to the very end.
Will the day's journey take the whole long day?
From morn to night, my friend.

But is there for the night a resting-place?
A roof for when the slow dark hours begin.
May not the darkness hide it from my face?
You cannot miss that inn.

Shall I meet other wayfarers at night?
Those who have gone before.
Then must I knock, or call when just in sight?
They will not keep you standing at that door.

Shall I find comfort, travel-sore and weary?
Of labour you shall find the sum.
Will there be beds for me and all who seek?
Yea, beds for all who come.

CHRISTINA ROSSETTI, 'Uphill'

I have come to the borders of sleep,
The unfathomable deep
Forest, where all must lose
Their way, however straight
Or winding, soon or late;
They can not choose.

Many a road and track
That since the dawn's first crack
Up to the forest brink
Deceived the travellers,
Suddenly now blurs,
And in they sink.

Here love ends—
Despair, ambition ends;
All pleasure and all trouble,
Although most sweet or bitter,
Here ends, in sleep that is sweeter
Than tasks most noble.

There is not any book
Or face of dearest look
That I would not turn from now
To go into the unknown
I must enter, and leave, alone,
I know not how.

The tall forest towers:
Its cloudy foliage lowers
Ahead, shelf above shelf:
Its silence I hear and obey
That I may lose my way
And myself.

EDWARD THOMAS (1878–1917), 'Lights Out'

Fear death?—to feel the fog in my throat,
The mist in my face,
When the snows begin, and the blasts denote
I am nearing the place,

The power of the night, the press of the storm,
    The post of the foe;
Where he stands, the Arch Fear in a visible form,
    Yet the strong man must go:
For the journey is done and the summit attained,
    And the barriers fall,
Though a battle's to fight ere the guerdon be gained,
    The reward of it all.
I was ever a fighter, so—one fight more,
    The best and the last!
I would hate that death bandaged my eyes, and forbore,
    And bade me creep past.
No! let me taste the whole of it, fare like my peers
    The heroes of old,
Bear the brunt, in a minute pay glad life's arrears
    Of pain, darkness and cold.
For sudden the worst turns the best to the brave,
    The black minute's at end,
And the elements' rage, the fiend-voices that rave,
    Shall dwindle, shall blend,
Shall change, shall become first a peace out of pain,
    Then a light, then thy breast,
O thou soul of my soul! I shall clasp thee again,
    And with God be the rest!

ROBERT BROWNING (1812–89), 'Prospice'

To my question, whether we might not fortify our minds for the approach of death, he answered, in a passion, 'No, Sir, let it alone. It matters not how a man dies, but how he lives. The act of dying is not of importance, it lasts so short a time.'

JAMES BOSWELL, *Life of Johnson*, 1791 (Samuel Johnson, 1709–84)

Death. On thee thou must take a long journey;
  Therefore thy book of count with thee thou bring,
  For turn again thou cannot by no way.
  And look thou be sure of thy reckoning,
  For before God thou shalt answer, and show
  Thy many bad deeds, and good but a few;
  How thou hast spent thy life, and in what wise,
  Before the chief Lord of paradise.
  Have ado that we were in that way,
  For, wit thou well, thou shalt make none attorney.

Everyman. Full unready I am such reckoning to give.
  I know thee not. What messenger art thou?
Death. I am Death, that no man dreadeth,
  For every man I rest, and no man spareth;
  For it is God's commandment
  That all to me should be obedient.
Everyman. O Death, thou comest when I had thee least in mind!
  In thy power it lieth me to save;
  Yet of my good will I give thee, if thou will be kind—
  Yea, a thousand pound shalt thou have—
  And defer this matter till another day.
Death. Everyman, it may not be, by no way.
  I set not by gold, silver, nor riches,
  Ne by pope, emperor, king, duke, ne princes;
  For, and I would receive gifts great,
  All the world I might get;
  But my custom is clean contrary.
  I give thee no respite. Come hence, and not tarry.

Everyman, *c.*1500

'Soon I must drink the poison; and I think that I had better repair to the bath first, in order that the women may not have the trouble of washing my body after I am dead.'...

'We will do our best,' said Crito: 'And in what way shall we bury you?'

'In any way that you like; but you must get hold of me, and take care that I do not run away from you.' Then he turned to us, and added with a smile: 'I cannot make Crito believe that I am the same Socrates who have been talking and conducting the argument; he fancies that I am the other Socrates whom he will soon see, a dead body—and he asks, How shall he bury me? . . . Be of good cheer then, my dear Crito, and say that you are burying my body only, and do with that whatever is usual, and what you think best.' . . .

Hitherto most of us had been able to control our sorrow; but now when we saw him drinking, and saw too that he had finished the draught, we could no longer forbear, and in spite of myself my own tears were flowing fast; so that I covered my face and wept, not for him, but at the thought of my own calamity in having to part from such a friend. Nor was I the first; for Crito, when he found himself unable to restrain his tears, had got up, and I followed; and at that moment Apollodorus, who had been weeping all the time, broke out in a loud and passionate cry which made cowards of us all.

Socrates alone retained his calmness: 'What is this strange outcry?' he said. 'I sent away the women mainly in order that they might not misbehave in this way, for I have been told that a man should die in peace. Be quiet then, and have patience.' . . .

He was beginning to feel cold about the groin, when he uncovered his face, for he had covered himself up, and said—they were his last words—he said: 'Crito, I owe a cock to Asclepius; will you remember to pay the debt?'

'The debt shall be paid,' said Crito: 'Is there anything else?' There was no answer to this question.

PLATO, *Phaedo*, tr. Benjamin Jowett

Re-enter Lear, with Cordelia dead in his arms.

Lear. Howl, howl, howl! O! you are men of stones:
Had I your tongues and eyes, I'd use them so
That heaven's vault should crack. She's gone for ever.
I know when one is dead, and when one lives;
She's dead as earth. Lend me a looking-glass;
If that her breath will mist or stain the stone,
Why, then she lives . . .
This feather stirs; she lives! If it be so,
It is a chance which does redeem all sorrows
That ever I have felt.
Kent. [*Kneeling*] O my good master!
Lear. Prithee, away.
Edgar. 'Tis noble Kent, your friend.
Lear. A plague upon you, murderers, traitors all!
I might have sav'd her; now she's gone for ever!
Cordelia, Cordelia! stay a little. Ha!
What is't thou say'st? Her voice was ever soft,
Gentle and low, an excellent thing in woman.
I kill'd the slave that was a-hanging thee.
Officer. 'Tis true, my lords, he did.
Lear. Did I not, fellow?
I have seen the day, with my good biting falchion
I would have made them skip: I am old now,
And these same crosses spoil me. Who are you?
Mine eyes are not o' th' best: I'll tell you straight . . .
And my poor fool is hang'd! No, no, no life!
Why should a dog, a horse, a rat, have life,
And thou no breath at all? Thou'lt come no more,
Never, never, never, never, never!
Pray you, undo this button: thank you, Sir.

Do you see this? Look on her, look, her lips,
Look there, look there! [*Dies*]
Edg. He faints! My Lord, my Lord!
Kent. Break, heart; I prithee, break!
Edg. Look up, my Lord.
Kent. Vex not his ghost: O! let him pass; he hates him
That would upon the rack of this tough world
Stretch him out longer.

SHAKESPEARE, *King Lear*

On the twenty-ninth day, being the day before he departed, he called for Doña Ximena, and for the Bishop Don Hieronymo, and Don Alvar Fañez Minaya, and Pero Bermudez, and his trusty Gil Diaz; and when they were all five before him, he began to direct them what they should do after his death; and he said to them, Ye know that King Bucar will presently be here to besiege this city, with seven and thirty Kings whom he bringeth with him, and with a mighty power of Moors. Now therefore the first thing which ye do after I have departed, wash my body with rose-water many times and well, as blessed be the name of God it is washed within and made pure of all uncleanness to receive his holy body tomorrow, which will be my last day. And when it has been well washed and made clean, ye shall dry it well, and anoint it with this myrrh and balsam, from these golden caskets, from head to foot, so that every part shall be anointed, till none be left. And you my Sister Doña Ximena, and your women, see that ye utter no cries, neither make any lamentation for me, that the Moors may not know of my death. And when the day shall come in which King Bucar arrives, order all the people of Valencia to go upon the walls, and sound your trumpets and tambours, and make the greatest rejoicings that ye can. And when ye would set out for Castille, let all the people know in secret, that they make themselves ready, and take with them all that they have, so that none of the Moors in the suburb may know thereof; for certes ye cannot keep the city, neither abide therein after my death. And see ye that sumpter beasts be laden with all that there is in Valencia, so that nothing which can profit may be left. And this I leave especially to your charge, Gil Diaz. Then saddle ye my horse Bavieca, and arm him well; and ye shall apparel my body full seemlily, and place me upon the horse, and fasten and tie me thereon so that it cannot fall: and fasten my sword Tizona in my hand. And let the Bishop Don Hieronymo go on one side of me, and my trusty Gil Diaz on the other, and he shall lead my horse. You, Pero Bermudez, shall bear my banner, as you were wont to bear it; and you, Alvar Fañez, my cousin, gather your company together, and put the host in order as you are wont to do. And go ye forth and fight with King Bucar: for be ye certain and doubt not that ye shall win this battle; God hath granted me this.

The Chronicle of the Cid (?12th century), tr. Robert Southey, 1808

With B.E.F. June 10. Dear Wife,
(O blast this pencil. 'Ere, Bill, lend's a knife.)
I'm in the pink at present, dear.
I think the war will end this year.
We don't see much of them square-'eaded 'Uns.
We're out of harm's way, not bad fed.
I'm longing for a taste of your old buns.
(Say, Jimmie, spare's a bite of bread.)
There don't seem much to say just now.
(Yer what? Then don't, yer ruddy cow!
And give us back me cigarette!)
I'll soon be 'ome. You mustn't fret.
My feet's improvin', as I told you of.
We're out in rest now. Never fear.
(VRACH! By crumbs, but that was near.)
Mother might spare you half a sov.
Kiss Nell and Bert. When me and you—
(Eh! What the 'ell! Stand to? Stand to!
Jim, give's a hand with pack on, lad.
Guh! Christ! I'm hit. Take 'old. Aye, bad.
No, damn your iodine. Jim? 'Ere!
Write my old girl, Jim, there's a dear.)

WILFRED OWEN (1893–1918), 'The Letter'

'He knows perfectly well—an old-timer like him—' I went on, 'that within an hour or two he will go to the gas chamber, naked, without his shirt, and without his package. What an extraordinary attachment to the last bit of property! After all, he could have given it to someone. I know that I'd never . . .'

'You think so, yes?' said the doctor indifferently. He took his hand off my shoulder, his jaws working as if he were sucking at a bad tooth.

'Forgive me, Doctor, but I feel certain that you too . . .' I added.

The doctor came from Berlin, had a daughter and a wife in Argentina, and he would sometimes speak of himself as *wir Preussen*, with a smile that combined the bitterness of a Jew with the pride of a former Prussian officer.

'I don't know. I don't know what I would do if I were going to the gas chamber. I might also want to take along my package.'

He turned towards me with a shy smile. I noticed that he was very tired and looked as if he had not slept for days.

'I think that even if I was being led to the oven, I would still believe that something would surely happen along the way. Holding a package would be a little like holding somebody's hand, you see.'

TADEUSZ BOROWSKI (1922–51), 'Auschwitz, Our Home: A Letter', tr. Barbara Vedder

Look upon Saint Lawrence, lying broiling upon the burning coals, as merry and as quiet as though he lay upon sweet red roses. When the tormentors turned his body upon the fiery gridirons, he bade the cruel tyrant eat of his burned side, whilst the other part was a-roasting. This saying declared that this holy martyr feared no death.

LUPSET, *The Way of Dying Well*

And so was he brought by Mr Lieutenant out of the Tower, and from thence led towards the place of execution, where going up the scaffold, which was so weak that it was ready to fall, he said to Mr Lieutenant, 'I pray you, I pray you, Mr Lieutenant, see me safe up, and for my coming down let me shift for myself.' Then desired he all the people thereabouts to pray for him, and to bear witness with him, that he should then suffer death in and for the faith of the holy Catholic Church, which done he kneeled down, and after his prayers said, he turned to the executioner, and with a cheerful countenance spake unto him, 'Pluck up thy spirits, man, and be not afraid to do thine office, my neck is very short. Take heed therefore thou shoot not awry for saving thine honesty.' So passed Sir Thomas More out of this world to God upon the very same day in which himself had most desired.

The execution of Sir Thomas More, St Thomas's Eve, 1535; from the *Life* by his son-in-law, William Roper

Where, twining subtile fears with hope,
He wove a net of such a scope,
That Charles himself might chase
To Caresbrooke's narrow case;

That thence the Royal actor borne
The tragic scaffold might adorn:
While round the armèd bands
Did clap their bloody hands.

He nothing common did or mean
Upon that memorable scene,
  But with his keener eye
  The axe's edge did try;

Nor call'd the gods, with vulgar spite,
To vindicate his helpless right;
  But bow'd his comely head
  Down, as upon a bed.

ANDREW MARVELL (1621–78), from 'An Horatian Ode upon Cromwell's Return from Ireland' (the execution of Charles I)

Duchess. What death?
Bosola. Strangling: here are your executioners.
Duch. I forgive them:
  The apoplexy, catarrh, or cough o'th' lungs
  Would do as much as they do.
Bos. Doth not death fright you?
Duch. Who would be afraid on't?
  Knowing to meet such excellent company
  In th'other world.
Bos. Yet, methinks,
  The manner of your death should much afflict you,
  This cord should terrify you?
Duch. Not a whit:
  What would it pleasure me to have my throat cut
  With diamonds? or to be smothered
  With cassia? or to be shot to death with pearls?
  I know death hath ten thousand several doors
  For men to take their exits; and 'tis found
  They go on such strange geometrical hinges,
  You may open them both ways:—any way, for heaven-sake,
  So I were out of your whispering:—tell my brothers
  That I perceive death, now I am well awake,
  Best gift is they can give, or I can take.
  I would fain put off my last woman's fault,
  I'd not be tedious to you.
Executioners. We are ready.
Duch. Dispose my breath how please you, but my body
  Bestow upon my women, will you?
Execut. Yes.
Duch. Pull, and pull strongly, for your able strength
  Must pull down heaven upon me:—

Yet stay; heaven-gates are not so highly arch'd
As princes' palaces, they that enter there
Must go upon their knees.

JOHN WEBSTER (*c.*1580–*c.*1625), *The Duchess of Malfi*

'Well,' said I, 'if it must be so, you have at least the satisfaction of leaving all your friends, your brother's family in particular, in great prosperity.' He said that he felt that satisfaction so sensibly, that when he was reading, a few days before, Lucian's *Dialogues of the Dead*, among all the excuses which are alleged to Charon for not entering readily into his boat, he could not find one that fitted him: he had no house to finish, he had no daughter to provide for, he had no enemies upon whom he wished to revenge himself. 'I could not well imagine,' said he, 'what excuse I could make to Charon, in order to obtain a little delay. I have done everything of consequence which I ever meant to do, and I could at no time expect to leave my relations and friends in a better situation than that in which I am now likely to leave them: I therefore have all reason to die contented.' He then diverted himself with inventing several jocular excuses, which he supposed he might make to Charon, and with imagining the very surly answers which it might suit the character of Charon to return to them. 'Upon further consideration,' said he, 'I thought I might say to him, "Good Charon, I have been correcting my works for a new edition. Allow me a little time that I may see how the public received the alterations." But Charon would answer, "When you have seen the effect of these, you will be for making other alterations. There will be no end of such excuses; so, honest friend, please step into the boat." But I might still urge, "Have a little patience, good Charon, I have been endeavouring to open the eyes of the public. If I live a few years longer, I may have the satisfaction of seeing the downfall of some of the prevailing systems of superstition." But Charon would then lose all temper and decency. "You loitering rogue, that will not happen these many hundred years. Do you fancy I will grant you a lease for so long a term? Get into the boat this instant, you lazy loitering rogue." '

ADAM SMITH, letter to William Strahan, 9 November 1776, on the death of David Hume

March 17 [1782] . . . Having heard that Thos. Thurston's Wife (who is and has been ill a long while) longed for some roast Veal from my House, having therefore a Loin roasted for Dinner, I sent her a good Plate of it.

March 21 . . . The poor Woman whom I sent some Veal to Sunday died yesterday morning—She eat nothing afterwards till she died, But she eat hearty of the Veal I sent her.

March 22 . . . I buried Eliz: Thurston Wife of Thos. Thurston this afternoon at Weston, aged 45 Yrs. It snowed all the whole Day with very cold high Wind.

December 12 [1798] . . . The first thing I heard this Morning when I came down Stairs, was the Death of my poor Clerk Thos. Thurston. He was out and at work on Monday at Mr Custance's. His Death was occasioned by a sudden & rapid Swelling in his Throat which suffocated him, something of the Quinsy. His Death was very sudden indeed . . . He was an harmless, industrious working Man as any in the Parish and very serviceable. Dinner today a fresh Tongue boiled & a Duck & c. Bitter cold today with rough NEE Wind which pinches me in the extreme. Willm. Large applied for the Clerkship this morning & I appointed him to the same.

JAMES WOODFORDE, *The Diary of a Country Parson, 1758–1802*

Bardolph. Would I were with him, wheresome'er he is, either in heaven or in hell!

Hostess. Nay, sure, he's not in hell: he's in Arthur's bosom, if ever man went to Arthur's bosom. A' made a finer end, and went away an it had been any christom child; a' parted ev'n just between twelve and one, ev'n at the turning o' th' tide: for after I saw him fumble with the sheets and play with flowers and smile upon his fingers' end, I knew there was but one way; for his nose was as sharp as a pen, and a' babbled of green fields. 'How now, Sir John?' quoth I: 'what, man! be o' good cheer.' So a' cried out 'God, God, God!' three or four times: now I, to comfort him, bid him a' should not think of God, I hoped there was no need to trouble himself with any such thoughts yet. So a' bade me lay more clothes on his feet: I put my hand into the bed and felt them, and they were as cold as any stone; then I felt to his knees, and so upward, and upward, and all was as cold as any stone.

Nym. They say he cried out of sack.

Host. Ay, that a' did.

Bard. And of women.

Host. Nay, that a' did not.

Boy. Yes, that a' did; and said they were devils incarnate.

Host. A' could never abide carnation; 'twas a colour he never liked.

Boy. A' said once, the devil would have him about women.
Host. A' did in some sort, indeed, handle women; but then he was rheumatic, and talked of the whore of Babylon.

SHAKESPEARE, *Henry V* (the death of Falstaff)

The most celebrated old man in seventeenth-century England was Thomas Parr of Winnington in Shropshire, turned into a sideshow in 1635 by the Earl of Arundel, the theme of a poem, 'The Old, Old, Very Old Man', by John Taylor the 'Water Poet', and the subject of an autopsy following his death at the age of 152 (or so it was said) performed by William Harvey, the famous physician. Parr had led an anonymous and uneventful life until he married for the first time when he was eighty. Twenty-five years later his affections were caught wandering, and he was compelled to stand in church clad in a white sheet as a penance for adultery. A second marriage followed at 112, to the stated satisfaction of his new wife. The Earl of Arundel brought Parr to London as a curiosity forty years later, where he promptly succumbed to the 'smoke of sulphurous coal constantly used as fuel for fires' and the more 'generous rich and varied diet, and stronger drink' than had been available in Shropshire. He was interred in Westminster Abbey.

LESLIE CLARKSON, *Death, Disease and Famine in Pre-Industrial England*, 1975

When the parish priest came to administer the last rites, Grandet's eyes, apparently dead for several hours past, brightened at the sight of the cross, the candlesticks, the silver stoup, which he stared at fixedly, the wen on his nose twitching for the last time. As the priest held the silver-gilt crucifix to his lips for him to kiss the image of Christ, he moved as if to seize it, and this frightful gesture cost him his life. He called to Eugénie, whom he did not see, although she was kneeling beside him, her tears bathing a hand already cold.

'Bless me, father,' she said.

'Take good care of everything! You'll render full account to me, over there,' he said, proving by those last words that Christianity is a fit religion for misers.

HONORÉ DE BALZAC (1799–1850), *Eugénie Grandet*

Here was a man who now for the first time found himself looking into the eyes of death—who was passing through one of those rare moments of experience when we feel the truth of a commonplace, which is as different from what we call knowing it, as the vision of waters upon the earth is different from the delirious vision of the water which cannot be had to cool the burning tongue. When the commonplace 'We must all die' transforms itself suddenly into the acute consciousness 'I must die—and soon', then death grapples us, and his fingers are cruel; afterwards, he may come to fold us in his arms as our mother did, and our last moment of dim earthly discerning may be like the first. To Mr Casaubon now, it was as if he suddenly found himself on the dark river-brink and heard the plash of the oncoming oar, not discerning the forms, but expecting the summons. In such an hour the mind does not change its lifelong bias, but carries it onwards in imagination to the other side of death, gazing backward —perhaps with the divine calm of beneficence, perhaps with the petty anxieties of self-assertion. What was Mr Casaubon's bias his acts will give us a clue to. He held himself to be, with some private scholarly reservations, a believing Christian, as to estimates of the present and hopes of the future. But what we strive to gratify, though we may call it a distant hope, is an immediate desire; the future estate for which men drudge up city alleys exists already in their imagination and love. And Mr Casaubon's immediate desire was not for divine communion and light divested of earthly conditions; his passionate longings, poor man, clung low and mist-like in very shady places.

GEORGE ELIOT (1819–80), *Middlemarch*

His sufferings, growing more and more severe, did their work and prepared him for death . . . Hitherto each individual desire aroused by suffering or privation, such as hunger, fatigue, thirst, had brought enjoyment when gratified. But now privation and suffering were not followed by relief, and the effort to obtain relief only occasioned fresh suffering. And so all desires were merged in one—the desire to be rid of all this pain and from its source, the body. But he had no words to express this desire for deliverance, and so he did not speak of it, but from force of habit asked for the things that had once given him comfort. 'Turn me over on the other side,' he would say, and immediately after ask to be put back again. 'Give me some beef tea. Take away the beef tea. Talk of something: why are you all so silent?' And directly they began to talk, he would close his eyes, and would show weariness, indifference, and loathing . . .

While the priest was reading the prayers, the dying man showed no sign of life. His eyes were closed . . . When he had come to the end of the prayer, the priest put the cross to the cold forehead, then slowly wrapped it in his stole and, after standing in silence for a minute or two, touched the huge, bloodless hand that was turning cold.

'He is gone,' said the priest, and made to move away; but suddenly there was a faint stir in the clammy moustaches of the dying man, and from the depths of his chest came the words, sharp and distinct in the stillness:

'Not quite . . . Soon.'

A moment later the face brightened, a smile appeared under the moustaches, and the women who had gathered round began carefully laying out the corpse.

LEO TOLSTOY (1828–1910), *Anna Karenina*, tr. Rosemary Edmonds (the death of Nikolai Levin)

Contrary to all expectation, he lived over the night; but *suffered much*, as well from his *impatience* and *disappointment* as from his *wounds*; for he seemed *very unwilling to die*.

He was delirious at times in the two last hours; and then several times cried out, as if he had seen some frightful spectre, Take her away! take her away! but named nobody. And sometimes praised some lady (that Clarissa, I suppose, whom he had invoked when he received his death's wound), calling her, Sweet Excellence! Divine Creature! Fair Sufferer! And once he said, Look down, Blessed Spirit, look down—And there stopped; his lips, however, moving.

At nine in the morning he was seized with convulsions, and fainted away; and it was a quarter of an hour before he came out of them.

His few last words I must not omit, as they show an ultimate composure; which may administer some consolation to his honourable friends.

Blessed—said he, addressing himself no doubt to Heaven; for his dying eyes were lifted up. A strong convulsion prevented him for a few moments saying more, but recovering, he again, with great fervour (lifting up his eyes and his spread hands), pronounced the word *blessed*. Then, in a seeming ejaculation, he spoke inwardly, so as not to be understood: at last, he distinctly pronounced these three words,

LET THIS EXPIATE!

And then, his head sinking on his pillow, he expired, at about half an hour after ten.

SAMUEL RICHARDSON (1689–1761), *Clarissa Harlowe* (the death of Lovelace)

If she avoided looking in the direction of her reposing husband it was not because she was afraid of him. Mr Verloc was not frightful to behold. He looked comfortable. Moreover, he was dead. Mrs Verloc entertained no vain delusions on the subject of the dead. Nothing brings them back, neither love nor hate. They can do nothing to you. They are as nothing. Her mental state was tinged by a sort of austere contempt for that man who had let himself be killed so easily. He had been the master of a house, the husband of a woman, and the murderer of her Stevie. And now he was of no account in every respect. He was of less practical account than the clothing on his body, than his overcoat, than his boots—than that hat lying on the floor. He was nothing. He was not worth looking at. He was even no longer the murderer of poor Stevie. The only murderer that would be found in the room when people came to look for Mr Verloc would be—herself!

JOSEPH CONRAD (1857–1924), *The Secret Agent*

Nobody could believe that it was true. Dr Marjoribanks's patients waited for him, and declared to their nurses that it was all a made-up story, and that he would come and prove that he was not dead. How could he be dead? He had been as well as he ever was that last evening. He had gone down Grange Lane in the snow, to see the poor old lady who was now sobbing in her bed, and saying it was all a mistake, and that it was she who ought to have died. But all those protestations were of no avail against the cold and stony fact which had frightened Thomas out of his senses, when he went to call the Doctor. He had died in the night without calling or disturbing anybody. He must have felt faint, it seemed, for he had got up and taken a little brandy, the remains of which still stood on the table by his bedside; but that was all that anybody could tell about it. They brought Dr Rider, of course; but all that he could do was to examine the strong, still frame —old, and yet not old enough to be weakly, or to explain such sudden extinction—which had ceased its human functions. And then the news swept over Carlingford like a breath of wind, though there was no wind even on that silent snowy day to carry the matter. Dr Marjoribanks was dead. It put the election out of people's heads, and even their own affairs for the time being; for had he not known all about the greater part of them—seen them come into the world and kept them in it—and put himself always in the breach when the pale Death approached that way? He had never made very much boast of his friendliness or been large in sympathetic expressions, but yet he had never flinched at any time, or deserted his patients for any consideration. Carlingford was sorry, profoundly sorry, with that true sorrow which is not so much for the person mourned as for the mourner's self, who feels a sense of something lost. The people said to themselves, Whom could they ever find who would know their constitutions so well, and who was to take care of So-and-so if he had another attack? To be sure Dr Rider was at hand, who felt a little agitated about it, and was conscious of the wonderful opening, and was very ready to answer, 'I am here'; but a young doctor is different from an old one, and a living man all in commonplace health and comfort is not to be compared with a dead one, on the morning at least of his sudden ending. Thank Heaven, when a life is ended there is always that hour or two remaining to set straight the defective balances and do a hasty late justice to the dead, before the wave sweeps on over him and washes out the traces of his steps, and lets in the common crowd to make their thoroughfare over the grave.

. . . to tell the truth, Dr Marjoribanks was one of the men who, according to external appearance, need never have died. There was nothing about him that wanted to be set right, no sort of loss, or failure, or misunderstanding, so far as anybody could see. An existence in which he could have his friends to dinner every week, and a good house, and good wine, and a very good table, and nothing particular to put him out of his way, seemed in fact the very ideal of the best life for the Doctor. There was nothing in him that seemed to demand anything better, and it was confusing to try to follow him into that which, no doubt, must be in all its fundamentals a very different kind of world. He was a just man and a good man in his way, and had been kind to many people in his lifetime—but still he did not seem to have that need of another rectifying, completer existence which most men have. There seemed no reason why he should die—a man who was so well contented with this lower region in which many of us fare badly, and where so few of us are contented. This was a fact which exercised a very confusing influence, even when they themselves were not aware of it, on many people's minds. It was hard to think of him under any other circumstances, or identify him with angels and spirits—which feeling on the whole made the regret for him a more poignant sort of regret.

And they buried him with the greatest signs of respect. People from twenty miles off sent their carriages, and all the George Street people shut their shops, and there was very little business done all day.

MARGARET OLIPHANT (1828–97), *Miss Marjoribanks*

We watched her breathing thro' the night,
Her breathing soft and low,
As in her breast the wave of life
Kept heaving to and fro.

So silently we seem'd to speak—
So slowly mov'd about,
As we had lent her half our powers
To eke her living out.

Our very hopes belied our fears,
Our fears our hopes belied—
We thought her dying when she slept.
And sleeping when she died.

For when the morn came dim and sad—
And chill with early showers,
Her quiet eyelids clos'd—she had
Another morn than ours.

THOMAS HOOD (1799–1845), 'The Death-Bed'

How did she leave the world? with what contempt?
Just as she in it liv'd, and so exempt
From all affection! When they urg'd the cure
Of her disease, how did her soul assure
Her suff'rings, as the body had been away!
And to the torturers (her doctors) say,
Stick on your cupping-glasses, fear not, put
Your hottest caustics to, burn, lance, or cut:
'Tis but a body which you can torment,
And I, into the world, all soul, was sent!
Then comforted her lord, and blest her son;
Cheer'd her fair sisters in her race to run;
With gladness temper'd her sad parents' tears;
Made her friends joys, to get above their fears.
And, in her last act, taught the standers-by,
With admiration and applause, to die!

BEN JONSON (1572–1637), from 'An Elegy on the Lady Jane Paulet'

I heard a Fly buzz—when I died—
The Stillness in the Room
Was like the Stillness in the Air—
Between the Heaves of Storm—

The Eyes around—had wrung them dry—
And Breaths were gathering firm
For that last Onset—when the King
Be witnessed—in the Room—

I willed my Keepsakes—Signed away
What portion of me be
Assignable—and then it was
There interposed a Fly—

With Blue—uncertain stumbling Buzz—
Between the light—and me—
And then the Windows failed—and then
I could not see to see—

EMILY DICKINSON, 'I heard a Fly buzz'

The curtain falls; the play is done.
The ladies and the gents have gone.
And did they like it? Well, some paws,
I think, beat out some gloved applause.
A very worthy audience stood
And clapped its bard, with gratitude.
But now the building has gone dumb:
Light and delight give way to gloom.
Though listen: a mean wailing thud
Near the bare stage, in some dark place—
Perhaps the parting of a dud
String on an ancient double-bass—
And an ill-tempered rustling—that's
The stalls being searched by theatre rats.
It smells of oil-fumes, noisomely.
The last light moans and flaps about,
Desperately fizzing, and goes out.
That poor light was the last of me.

HEINRICH HEINE (1797–1856), 'It's going out', tr. Alistair Elliot

In Prague, a rabbi, name of Pelf,
Through Satan's magic came to gain
Such powers that even Death himself
Belched flames against him, but in vain.

In spite of all, though, Death won out:
He found a rose and hid inside.
Satan had not thought about
A rose; Pelf breathed her once, and died.

CHRISTIAN MORGENSTERN (1871–1914), 'The Rabbi', tr. W. D. Snodgrass and Lore Segal

When I put off the sense in death,
And lose all seeming with my breath,
I will not heed the prejudice of nose,
Comparisons of carrion and rose.

When this now fettered judgement shall be free,
All changes are of equal worth to me.
And I will pleasure in the faultless way
My flesh dissolves to worms and fertile clay.

ANNA WICKHAM (1884–1947), 'The Free Intelligence'

Fowler and Millsom, the Muswell Hill murderers, were executed together; but as Fowler, in the dock, had made a violent attempt to get at Millsom, who had turned Queen's evidence, they were kept apart upon the scaffold, and a man, Seaman, who was to be executed with them, was placed on the trap-door between them. At this, the penultimate moment of his life, a quaint conceit would seem to have been born in the mind of Seaman, who was heard to say: 'This is the first time as ever I was a —— peacemaker.'

ROBERT W. MACKENNA, *The Adventure of Death*, 1916

I am ready to meet my Maker. Whether my Maker is prepared for the ordeal of meeting me is another matter.

WINSTON CHURCHILL on his 75th birthday

Whoso maintains that I am humbled now
(Who wait the Awful Day) is still a liar;
I hope to meet my Maker brow to brow
And find my own the higher.

FRANCES CORNFORD (1886–1960), 'Epitaph for a Reviewer'

Oh the sad day,
When friends shall shake their heads and say
Of miserable me,
Hark how he groans, look how he pants for breath,
See how he struggles with the pangs of death!
When they shall say of these poor eyes,
How hollow, and how dim they be!
Mark how his breast does swell and rise,
Against his potent enemy!
When some old friend shall step to my bedside,
Touch my chill face, and thence shall gently slide,
And when his next companions say,
How does he do? What hopes? shall turn away,
Answering only with a lift-up hand,
Who can his fate withstand?
Then shall a gasp or two do more
Than e'er my rhetoric could before,
Persuade the peevish world to trouble me no more!

THOMAS FLATMAN (1637–88), 'Death'

Whoever coined the phrase *The Body Politic*?
All States we've lived in, or historians tell of,
have had shocking health, psychosomatic cases,
physicked by sadists or glozing expensive quacks:
when I read the papers, You seem an Adonis.

Time, we both know, will decay You, and already
I'm scared of our divorce: I've seen some horrid ones.
Remember: when *Le Bon Dieu* says to You *Leave him!*,
please, please, for His sake and mine, pay no attention
to my piteous *Don'ts*, but bugger off quickly.

W. H. AUDEN (1907–73), 'Talking to Myself'

Was I in a condition to stipulate with Death, as I am this moment with my apothecary, how and where I will take his clyster—I should certainly declare against submitting to it before my friends; and therefore I never seriously think upon the mode and manner of this great catastrophe, which generally takes up and torments my thoughts as much as the catastrophe itself; but I constantly draw the curtain across it with this wish, that the Disposer of all things may so order it, that it happen not to me in my own house—but rather in some decent inn—

at home, I know it,—the concern of my friends, and the last services of wiping my brows, and smoothing my pillow, which the quivering hand of pale affection shall pay me, will so crucify my soul, that I shall die of a distemper which my physician is not aware of: but in an inn, the few cold offices I wanted, would be purchased with a few guineas, and paid me with an undisturbed but punctual attention . . .

LAURENCE STERNE (1713–68), *Tristram Shandy*

Moreover I now understood very well how one could carry with one, through all the years, deep in one's portfolio, the description of a dying hour. It need not even be an especially selected one; they all possess something almost rare. Can one not, for example, imagine somebody copying out the description of Felix Arvers' death? It took place in a hospital. He was dying with ease and tranquillity, and the sister, perhaps, thought he had gone further with it than he really had. In a very loud voice she called out an order, indicating where such and such was to be found. She was hardly an educated nun, and had never seen written the word 'corridor', which at the moment she could not avoid using; thus it happened that she said 'collidor', thinking it ought to be pronounced so. At that Arvers thrust death from him. He felt it necessary to put this right first. He became perfectly lucid and explained to her that it ought to be pronounced 'corridor'. Then he died. He was a poet and hated the approximate; or perhaps he was simply concerned with the truth; or it annoyed him to carry away this last impression that the world would go on so carelessly. That can no longer be decided. Only let no one think that he acted in a spirit of pedantry. Otherwise the same reproach would fall on the saintly Jean-de-Dieu, who sprang up in the midst of his dying and arrived just in time to cut down a man who had hanged himself in his garden, knowledge of whom had in some amazing fashion penetrated the inward tension of his agony. He, too, was concerned only with the truth.

RILKE, *The Notebook of Malte Laurids Brigge*, tr. John Linton

'I know what death is, I am an old retainer of his; and believe me, he's overrated. Almost nothing to him. Of course, all kinds of beastliness can happen beforehand—but it isn't fair to count those in, they are as living as life itself, and can just as well lead up to a cure. But about death—no one who came back from it could tell you anything, because we don't realize it. We come out of the dark and go into the dark again, and in between lie the experiences of our life. But the beginning and the end, birth and death, we do not experience; they have no subjective character, they fall entirely in the category of objective events, and that's that.'

Which was the Hofrat's way of administering consolation. We may hope that the reasonable Frau Ziemssen drew comfort therefrom; his assurances, at least, were in a very large degree justified by the event. Joachim, in these days, slept many hours, out of weakness, and probably dreamed of the flat-land and the service and whatever else was pleasant to him to dream . . .

At seven o'clock he died . . . He had sunk down in the bed and curtly ordered them to prop him up. While Frau Ziemssen, with her arm about his shoulders, tried to do so, he said hurriedly that he must write out an application for an extension of his leave and hand it in at once; and even while he said this, the 'short crossing' came to pass, as Hans Castorp, reverently watching in the light of the red-shaded table-lamp, quickly perceived. His gaze grew dim, the unconscious tension of the features relaxed, the strained and swollen look about the lips notably diminished; the beauty of early manhood visited once more our Joachim's quiet brow, and all was over.

THOMAS MANN (1875–1955), *The Magic Mountain*, tr. H. T. Lowe-Porter

Tancredi. Yes, much on the credit side came from Tancredi; that sympathy of his, all the more precious for being ironic; the aesthetic pleasure of watching him manoeuvre amid the shoals of life, the bantering affection whose touch was so right. Then dogs; Fufi, the fat pug of his childhood, the impetuous poodle Tom, confidant and friend, Speedy's gentle eyes, Bendicò's delicious nonsense, the caressing paws of Pop, the pointer at that moment searching for him under bushes and garden chairs and never to see him again; then a horse or two, these already more distant and detached. There were the first few hours of returns to Donnafugata, the sense of tradition and the perennial expressed in stone and water, of time congealed; a few care-free shoots, a cosy massacre or two of hares and partridges, some good laughs with Tumeo, a few minutes of compunction at the convent amid odours of must and confectionery. Anything else? Yes, there were other things: but these were only grains of gold mixed with earth: moments of satisfaction when he had made some biting reply to a fool, of content when he had realized that in Concetta's beauty and character was prolonged the true Salina strain; a few seconds of frenzied passion; the surprise of Arago's letter spontaneously congratulating him on the accuracy of his difficult calculations about Huxley's comet. And—why not?—the public thrill of being given a medal at the Sorbonne, the exquisite sensation of one or two fine silk cravats, the smell of some macerated leathers, the gay voluptuous air of a few women passed in the street, of one glimpsed even yesterday at the station of Catania, in a brown travelling dress and suede gloves, mingling amid the crowds and seeming to search for his exhausted face through the dirty compartment window. What a noise that crowd was making! 'Sandwiches!' '*Il Corriere dell'Isola*.' And then the panting of the tired breathless train . . . and that appalling sun as they arrived, those lying faces, the crashing cataracts . . .

In the growing dark he tried to count how much time he had really lived. His brain could not cope with the simple calculation any more; three months, three weeks, a total of six months, six by eight, eighty-four . . . forty-eight thousand . . . √840,000. He summed up. 'I'm seventy-three years old, and all in all I may have lived, really lived, a total of two . . . three at the most.' And the pains, the boredom, how long had they been? Useless to try and make himself count those; the whole of the rest; seventy years.

He felt his hand no longer being squeezed. Tancredi got up hurriedly and went out . . . Now it was not a river erupting over him but an ocean, tempestuous, all foam and raging white-flecked waves . . .

He must have had another stroke for suddenly he realized that he was lying stretched on the bed. Someone was feeling his pulse; from the window came the blinding implacable reflection of the sea; in the room could be heard a faint hiss; it was his own death-rattle, but he did not know it. Around him was a little crowd, a group of strangers staring at him with frightened expressions. Gradually he recognized them: Concetta, Francesco Paolo, Carolina, Tancredi, Fabrizietto. The person holding his pulse was Doctor Cataliotti; he tried to smile a greeting at the latter but no one seemed to notice; all were weeping except Concetta; even Tancredi, who was saying: 'Uncle, dearest Nuncle!'

Suddenly amid the group appeared a young woman; slim, in brown travelling dress and wide bustle, with a straw hat trimmed with a speckled veil which could not hide the sly charm of her face. She slid a little suede-gloved hand between one elbow and another of the weeping kneelers, apologized, drew closer. It was she, the creature for ever yearned for, coming to fetch him; strange that one so young should yield to him; the time for the train's departure must be very close. When she was face to face with him she raised her veil, and there, chaste but ready for possession, she looked lovelier than she ever had when glimpsed in stellar space.

The crashing of the sea subsided altogether.

GIUSEPPE DI LAMPEDUSA (1896–1957), *The Leopard*, tr. Archibald Colquhoun (the death of Prince Fabrizio)

When later, after a period when Manya seemed better, alarming symptoms developed and she was told that she had only a few weeks before her, she made no mention of this or of the fact that she had received the Last Sacraments, but ordered champagne all round. She continued working until she was too weak to do so. Even if she had not told a friend 'I am infinitely happy and only sad that I am too weak to share my happiness' it was plain that she was at peace. Nor did she lose her sense of humour. To someone who had been praying until recently for her recovery and had now changed her intention, but had certainly never made any mention of this, she remarked: 'Thank goodness you look quite different since you have stopped telling God what to do about me.'

When she first knew she was seriously ill, she had written to her son, 'dying is a job . . . the important thing is to do it as well as one can'. When death was imminent her attitude was the same.

Two days before she died, she looked for the first time troubled and said: 'Until now I have always known what God wanted me to do, but now I feel quite extraordinary, and I am not sure what I should do. Do you think the cancer has reached my brain?'

When told that she was actually dying and that it would not last long, she relaxed and said 'In that case it is very simple, but would it be peculiar if I asked to have the Last Sacraments again?'

Manya Harari: Memoirs 1906–69, 1972

Before 1959 when Herman Feifel wanted to interview the dying about themselves, no doubt for the first time, hospital authorities were indignant. They found the project 'cruel, sadistic, traumatic'. In 1965 when Elisabeth Kübler-Ross was looking for dying persons to interview, the heads of the hospitals and clinics to whom she addressed herself protested, 'Dying? But there are no dying here!' There could be no dying in a well-organized and respectable institution. They were mortally offended.

ARIÈS, *The Hour of Our Death*

On pillow after pillow lies
The wild white hair and staring eyes;
Jaws stand open; necks are stretched
With every tendon sharply sketched;
A bearded mouth talks silently
To someone no one else can see.

Sixty years ago they smiled
At lover, husband, first-born child.

Smiles are for youth. For old age come
Death's terror and delirium.

PHILIP LARKIN (b. 1922), 'Heads in the Women's Ward'

At the end of a long-walled garden
  in a red provincial town,
A brick path led to a mulberry
  scanty grass at its feet.
I lay under blackening branches
  where the mulberry leaves hung down
Sheltering ruby fruit globes
  from a Sunday-tea-time heat.
Apple and plum espaliers
  basked upon bricks of brown;
The air was swimming with insects,
  and children played in the street.

Out of this bright intentness
  into the mulberry shade
Musca domestica (housefly)
  swung from the August light
Slap into slithery rigging
  by the waiting spider made
Which spun the lithe elastic
  till the fly was shrouded tight.
Down came the hairy talons
  and horrible poison blade
And none of the garden noticed
  that fizzing, hopeless fight.

Say in what Cottage Hospital
  whose pale green walls resound
With the tap upon polished parquet
  of inflexible nurses' feet
Shall I myself be lying
  when they range the screens around?
And say shall I groan in dying
  as I twist the sweaty sheet?
Or gasp for breath uncrying,
  as I feel my senses drown'd
While the air is swimming with insects
  and children play in the street.

JOHN BETJEMAN (b. 1906), 'The Cottage Hospital'

It begins with an easy voice saying,
'Just a routine examination',
as October sunlight
pierces the heavy velvet curtains.

Later, it is the friends who write but do not visit,
it is (after all these years)
getting thinner,
it is boiled fish,
it is a folded screen in the corner,
it is doctors who no longer stop by your bed
but only say 'Good-night' or 'Good-morning';
it is terror every minute of conscious night and day
to a background of pop music.
It is trying not to think of it.

At the end you are left with the thought
that from Mozart to Siegfried Sassoon
all the people you respect
are dead anyhow.

LYALL WILKES (b. 1914), 'Nightmare'

As he came near death things grew shallower for us:
We'd lost sleep and now sat muffled in the scent of tulips, the medical odours, and the street sounds going past, going away;
And he, too, slept little, the morphine and the pink light the curtains let through floating him with us,
So that he lay and was worked out on to the skin of his life and left there,
And we had to reach only a little way into the warm bed to scoop him up.

A few days, slow tumbling escalators of visitors and cheques, and something like popularity;
During this time somebody washed him in a soap called *Narcissus* and mounted him, frilled with satin, in a polished case.

Then the hole: this was a slot punched in a square of plastic grass rug, a slot lined with white polythene, floored with dyed green gravel.
The box lay in it; we rode in the black cars round a corner, got out into our coloured cars and dispersed in easy stages.

After a time the grave got up and went away.

ROY FISHER (b. 1930), 'As He Came Near Death'

I am a student nurse. I am dying. I write this to you who are, and will become, nurses in the hope that by my sharing my feelings with you, you may someday be better able to help those who share my experience.

I'm out of the hospital now—perhaps for a month, for six months, perhaps for a year—but no one likes to talk about such things. In fact, no one likes to talk about much at all. Nursing must be advancing, but I wish it would hurry. We're taught not to be overly cheery now, to omit the 'Everything's fine' routine, and we have done pretty well. But now one is left in a lonely silent void. With the protective 'fine, fine' gone, the staff is left with only their own vulnerability and fear. The dying patient is not yet seen as a person and thus cannot be communicated with as such. He is a symbol of what every human fears and what we each know, at least academically, that we too must someday face. What did they say in psychiatric nursing about meeting pathology with pathology to the detriment of both patient and nurse? And there was a lot about knowing one's own feelings before you could help another with his. How true.

But for me, fear is today and dying is now. You slip in and out of my room, give me medications and check my blood pressure. Is it because I am a student nurse, myself, or just a human being, that I sense your fright? And your fears enhance mine. Why are you afraid? I am the one who is dying!

I know you feel insecure, don't know what to say, don't know what to do. But please believe me, if you care, you can't go wrong. Just admit that you care. That is really for what we search. We may ask for why's and wherefore's, but we don't really expect answers. Don't run away—wait—all I want to know is that there will be someone to hold my hand when I need it. I am afraid. Death may get to be a routine to you, but it is new to me. You may not see me as unique, but I've never died before. To me, once is pretty unique!

You whisper about my youth, but when one is dying, is he really so young anymore? I have lots I wish we could talk about. It really would not take much more of your time because you are in here quite a bit anyway.

If only we could be honest, both admit of our fears, touch one another. If you really care, would you lose so much of your valuable professionalism if you even cried with me? Just person to person? Then, it might not be so hard to die—in a hospital—with friends close by.

ANON., *American Journal of Nursing*, 1970

One day I visited Dr Horiuchi and inquired how many days and months before my death. I asked him to speak truthfully and hide nothing. With much to do and enjoy, I wanted to use completely every last day. To make plans for my remaining days I asked how long I had to live. The rather innocent Dr Horiuchi thought for a few minutes and then replied quite uncomfortably, 'One year and a half; perhaps two years if you take good care of yourself.' I told Horiuchi that I had expected to live only five or six months and that in one year I could certainly reap a rich harvest from life.

Some of you may say that one year and a half is very short; I say it is an eternity. But if you wish to say it is short, then ten years is also short, and fifty years is short, and so, too, is one hundred years. If this life is limited in time and that after death is unlimited, then the limited compared to the unlimited period is not even short: it's nothing. If you have things to do and enjoy, then isn't it possible to use quite well one year and a half? As fifty and one hundred years disappear, so, too, does the so-called one year and a half. Our life is nothing but a single empty boat on a non-existent sea.

NAKAE CHŌMIN (1847–1901), *One Year and a Half*, tr. Robert Jay Lifton, Shūichi Katō and Michael R. Reich

Richer now the body's juices,
The flavour of a lifetime.
I wash myself more often these days,
Drops gather in my navel—
I have more time for washing.

In the oyster-light of morning
I glance coolly at my stool
(Is the future there presaged
Or only yesterday remembered?).
I feel I am all soul.

I peer at some stained scroll;
A faithful friend, it peers at me.

All soul!—Now I have time to be.
Age suits me best of all my ages.

TAO TSCHUNG YU (?18th century)

The dying often find the settling of various practical matters concerning family life, property and work responsibility a satisfying task. They can get a sense of completion. They can take pleasure in ensuring that their dependents will be grateful for their forethought. Sometimes, those who are aware that their existence is to be curtailed, may get a little more from life by advancing their plans, even a proposed marriage, and so have further shared pleasures, bitter-sweet though they may be. It is not uncommon for dying people to get more pleasure out of their remaining days than others would believe possible. Couples who have married in spite of knowing that one of them was to die, may be very sad when the death does draw near and others grieve for them; but they are usually quite sure that they were right to take advantage of the life that could be enjoyed. Other people, when dying, will not want to *do* things but can quite pleasurably review their life, on the whole satisfied with its achievements and satisfactions. Some, knowing the situation, will wish to prepare themselves spiritually for their anticipated eternal existence.

JOHN HINTON, *Man's Concern With Death*, 1968

If I die,
leave the balcony open.

The little boy is eating oranges.
(From my balcony I can see him.)

The reaper is harvesting the wheat.
(From my balcony I can hear him.)

If I die,
leave the balcony open!

FEDERICO GARCÍA LORCA (1898–1936), 'Farewell', tr. W. S. Merwin

Christ.
May I die at night
With the semblance of my senses
Like the full moon that fails.

ROBERT LOWELL (1917–77) (found among his papers)

As Amr lay on his death-bed a friend said to him: 'You have often remarked that you would like to find an intelligent man at the point of death, and to ask him what his feelings were. Now I ask *you* that question.' Amr replied, 'I feel as if heaven lay close upon the earth and I between the two, breathing through the eye of a needle.'

AMR IBN AL-AS, the Arab conqueror of Egypt (d. 664)

Since I am coming to that holy room,
Where, with thy quire of Saints for evermore,
I shall be made thy music; as I come
I tune the Instrument here at the door,
And what I must do then, think here before.

Whilst my Physicians by their love are grown
Cosmographers, and I their Map, who lie
Flat on this bed, that by them may be shown
That this is my South-west discovery
Per fretum febris, by these straits to die,

I joy, that in these straits, I see my West;
For, though their currents yield return to none,
What shall my West hurt me? As West and East
In all flat Maps (and I am one) are one,
So death doth touch the Resurrection.

Is the Pacific Sea my home? Or are
The Eastern riches? Is Jerusalem?
Anyan, and Magellan, and Gibraltare,
All straits, and none but straits, are ways to them,
Whether where Japhet dwelt, or Cham, or Sem.

We think that Paradise and Calvary,
Christ's Cross, and Adam's tree, stood in one place;
Look, Lord, and find both Adams met in me;
As the first Adam's sweat surrounds my face,
May the last Adam's blood my soul embrace.

So, in his purple wrapp'd receive me, Lord,
By these his thorns give me his other Crown;
And as to others' souls I preach'd thy word,
Be this my Text, my Sermon to mine own:
Therefore that he may raise, the Lord throws down.

DONNE, 'Hymn to God my God, in my Sickness'

They have put my bed beside the unpainted screen;
They have shifted my stove in front of the blue curtain.
I listen to my grandchildren reading me a book;
I watch the servants heating up my soup.
With rapid pencil I answer the poems of friends,
I feel in my pockets and pull out medicine-money.
When this superintendence of trifling affairs is done,
I lie back on my pillows and sleep with my face to the South.

Po Chü-i (772–846), 'Last Poem', tr. Arthur Waley

Suicide

WE are on delicate ground here, as the letter in *The Times* apropos of EXIT makes clear. Although we respect the right to end one's life—a right that somehow seems more real than the right to happiness, full employment and so forth—we may not wish to see it exercised too readily, whether because of some indefinable aversion-working of natural instinct, or for one of the various reasons advanced by Voltaire (the man who kills himself in a melancholy fit today would have wanted to live had he waited a week), or because of the effects on the family of the suicide. In that last connection a poem of Emily Dickinson's, though not explicitly to do with suicide, is a gentle reminder:

> So proud she was to die
> It made us all ashamed
> That what we cherished, so unknown
> To her desire seemed—
> So satisfied to go
> Where none of us should be
> Immediately—that Anguish stooped
> Almost to Jealousy.

Does one have the right to impose such a burden, however indeterminate its implications, on other people? For many of us at times the idea of suicide has the kind of attraction indicated in Cleopatra's little speech, as an act we *choose* to perform. Pavese expounds this view lucidly and succinctly. At other times it will repel, as being merely a childish anticipation of what must in any case occur sooner or later. Paul-Louis Landsberg reckons that more people abstain from suicide out of cowardice than kill themselves because they are cowardly. Elsewhere in this book it is suggested that those who have most reason to end their lives—lives of terror and suffering—may yet reject suicide outright. Certainly, killing oneself to spite some unobservant Caesar, or in the spirit of 'now they'll be sorry' or 'now they'll take me seriously' is, as two pieces in this section intimate, likely to have the contrary effect, or no effect at all.

With most of us, most of the time, what prevails is our sense of the ridiculous, our dislike for the melodramatic. How can one take oneself so seriously; how can one presume to pre-empt that huge and all-embracing thing, the future; how should one dare to inconvenience the public so? 'Do I dare to eat a peach?', to begin with; and 'They will say: "But how his arms and legs are thin!" ' Perhaps it is good taste that holds us back, or modesty, or pride—or merely a sense of humour, which isn't to be despised . . . Or, of course, hope which (along with curiosity), whether or not it springs eternal, can last out our limited existences. Experience suggests that there is something in the folk wisdom that tells us, When one door closes, another opens.

Several writers testify to the benefits gained from the thought of suicide as an ever-present stand-by: people as diverse as David Hume, Nietzsche and Stevie Smith have seen it as a theoretical tower of strength, or something put aside against a rainy day. Simpler and more sustaining is Johnson's recipe: the safest antidote to sorrow is employment. He was referring to grief at others' deaths, adducing the example of soldiers and seamen among whom he found much kindness but little grief: 'Sorrow is a kind of rust of the soul . . . remedied by exercise and motion.' Voltaire has the same advice to offer (in this case addressed to editors and other literati) when considering the English propensity to *felo de se*, 'our island's shame', according to Robert Blair: always have something on the stocks. While Stendhal, pushing French logic several steps further (and yet still in line with witnesses whose seriousness is beyond dispute), recommends a drastic regime of persecution in order to provoke the instinct to survive and thus drive out suicidal intentions.

We know that there are circumstances in which none of these considerations and prophylactics will carry sufficient weight. Virginia Woolf's fear of madness and the inability to go on writing, the blackmail that thwarted Carla Mann's effort to achieve respectability, the mixture of old and young in little Jude's desperate reasoning . . . But the harrowing letter left by Richard and Bridget Smith is proof enough, and the more forcible in that the couple were professing Christians. 'Who knows how he may be tempted?' asks Robert Burton, in what has to be the final verdict. 'It is his case; it may be thine.'

~

When, in some dreadful and ghastly dream, we reach the moment of greatest horror, it awakes us; thereby banishing all the hideous shapes that were born of the night. And life is a dream: when the moment of greatest horror compels us to break it off, the same thing happens.

Suicide may also be regarded as an experiment—a question which man puts to Nature, trying to force her to an answer. The question is this: What change will death produce in a man's existence and in his insight into the nature of things? It is a clumsy experiment to make;

for it involves the destruction of the very consciousness which puts the question and awaits the answer.

ARTHUR SCHOPENHAUER (1788–1860), *Parerga and Paralipomena*, tr. T. Bailey Saunders

Death will necessarily come, from ordinary causes. It is inevitable, and one's whole life is a preparation for it, an event as natural as the fall of raindrops. I cannot resign myself to that thought. Why not seek death of one's own free will, asserting one's right to choose, giving it some significance? Instead of passively letting it happen? Why not?

Here's the reason. One always puts off the decision, feeling (or hoping) that one more day, one more hour of life, might also prove an opportunity of asserting our freedom of choice, which we should lose by seeking death. In short, because one thinks—and I speak for myself—that there is plenty of time. So the day of natural death comes, and we have missed the great opportunity of performing, *for a specific reason*, the most important act in life.

CESARE PAVESE (1908–50), *This Business of Living*, tr. A. E. Murch

There is but one truly serious philosophical problem and that is suicide. Judging whether life is or is not worth living amounts to answering the fundamental question of philosophy. All the rest—whether or not the world has three dimensions, whether the mind has nine or twelve categories—comes afterwards . . .

Suicide has never been dealt with except as a social phenomenon. On the contrary, we are concerned here, at the outset, with the relationship between individual thought and suicide. An act like this is prepared within the silence of the heart, as is a great work of art. The man himself is ignorant of it. One evening he pulls the trigger or jumps. Of an apartment-building manager who had killed himself I was told that he had lost his daughter five years before, that he had changed greatly since and that that experience had 'undermined' him. A more exact word cannot be imagined. Beginning to think is beginning to be undermined. Society has but little connection with such beginnings. The worm is in man's heart.

ALBERT CAMUS (1913–60), *The Myth of Sisyphus*, tr. Justin O'Brien

Then both ourselves and seed at once to free
From what we fear for both, let us make short,
Let us seek Death, or he not found, supply
With our own hands his office on ourselves;

Why stand we longer shivering under fears
That show no end but death, and have the power,
Of many ways to die the shortest choosing,
Destruction with destruction to destroy?

MILTON, *Paradise Lost* (Eve is speaking)

There are certainly far more people who do not kill themselves because they are too cowardly to do so, than those who kill themselves out of cowardice . . . The Christian religion, which condemns suicide as sin, considers it far more the sin of Lucifer than a banal cowardice . . . Suicide is something on its own. It seems to me to be a flight by which man hopes to recover Paradise Lost instead of trying to deserve Heaven.

PAUL-LOUIS LANDSBERG (1901–43), *The Experience of Death and The Moral Problem of Suicide*, tr. Cynthia Rowland

My desolation does begin to make
A better life. 'Tis paltry to be Caesar:
Not being Fortune, he's but Fortune's knave,
A minister of her will: and it is great
To do that thing that ends all other deeds,
Which shackles accidents and bolts up change,
Which sleeps, and never palates more the dung,
The beggar's nurse, and Caesar's.

SHAKESPEARE, *Antony and Cleopatra*

To be, or not to be, that is the question,
Whether 'tis nobler in the mind to suffer
The slings and arrows of outrageous fortune,
Or to take arms against a sea of troubles,
And by opposing, end them. To die, to sleep—
No more, and by a sleep to say we end
The heart-ache, and the thousand natural shocks
That flesh is heir to; 'tis a consummation
Devoutly to be wished. To die, to sleep!
To sleep, perchance to dream, ay there's the rub,
For in that sleep of death what dreams may come
When we have shuffled off this mortal coil
Must give us pause—there's the respect
That makes calamity of so long life:
For who would bear the whips and scorns of time,

Th'oppressor's wrong, the proud man's contumely,
The pangs of disprized love, the law's delay,
The insolence of office, and the spurns
That patient merit of th'unworthy takes,
When he himself might his quietus make
With a bare bodkin; who would fardels bear,
To grunt and sweat under a weary life,
But that the dread of something after death,
The undiscovered country, from whose bourn
No traveller returns, puzzles the will,
And makes us rather bear those ills we have,
Than fly to others that we know not of?

SHAKESPEARE, *Hamlet*

He scanned it—staggered—
Dropped the Loop
To Past or Period—
Caught helpless at a sense as if
His Mind were going blind—

Groped up, to see if God was there—
Groped backward at Himself
Caressed a Trigger absently
And wandered out of Life.

EMILY DICKINSON, 'He scanned it'

Who travels by the wearie wandring way,
To come unto his wishèd home in haste,
And meets a flood, that doth his passage stay,
Is not great grace to help him over past,
Or free his feet that in the mire stick fast? . . .

What if some little pain the passage have,
That makes frail flesh to fear the bitter wave?
Is not short pain well borne, that brings long ease,
And lays the soul to sleep in quiet grave?
Sleep after toil, port after stormy seas,
Ease after war, death after life does greatly please.

The knight much wondered at his sudden wit,
And said, The term of life is limited,
Nor may a man prolong nor shorten it;
The soldier may not move from watchful stead,

Nor leave his stand, until his Captain bid.
Who life did limit by almighty doom
(Quoth he) knows best the terms establishèd;
And he, that points the Sentinel his room,
Doth license him depart at sound of morning drum.

EDMUND SPENSER (?1552–99), *The Faerie Queene* (a debate between Despair and the Red Cross Knight)

There is a doctrine whispered in secret that man is a prisoner who has no right to open the door and run away; this is a great mystery which I do not quite understand. Yet I too believe that the gods are our guardians, and that we men are a possession of theirs . . . And if one of your own possessions, an ox or an ass, for example, took the liberty of putting himself out of the way when you had given no intimation of your wish that he should die, would you not be angry with him, and would you not punish him if you could? . . . Then, if we look at the matter thus, there may be reason in saying that a man should wait, and not take his own life until God summons him, as he is now summoning me.

PLATO, *Phaedo*, tr. Benjamin Jowett

Just as a landlord who has not received his rent pulls down the doors, removes the rafters, and fills up the well, so I seem to be driven out of this little body, when nature, which has let it to me, takes away one by one, eyes and ears, hands and feet. I will not therefore delay longer, but will cheerfully depart as from a banquet.

GAIUS RUFUS MUSONIUS (*c.*30–*c.*100)

The advocates for suicide tell us that it is quite permissible to quit our house when we are weary of it. Agreed—but most men would rather lie in a ramshackle house than sleep in the open fields.

VOLTAIRE, *Lettres Philosophiques sur les Anglais*

It's bad that it is so untidy, there is no denying that, for it bespatters one's friends morally as well as physically, taking them so much more into one's secret than they want to be taken. But how heroic to be able to suppress one's vanity to the extent of confessing that the game is too hard. The most comic and apparently the chief argument against it, is that because you were born without being consulted, you would be very sinful should you cut short your blissful career!

The Diary of Alice James, 5 August 1889

A hair, a fly, an insect is able to destroy this mighty being whose life is of such importance. Is it an absurdity to suppose that human prudence may lawfully dispose of what depends on such insignificant causes? It would be no crime in me to divert the *Nile* or *Danube* from its course, were I able to effect such purposes. Where then is the crime of turning a few ounces of blood from their natural channel?—Do you imagine that I repine at providence or curse my creation, because I go out of life, and put a period to a being, which, were it to continue, would render me miserable? Far be such sentiments from me; I am only convinced of a matter of fact, which you yourself acknowledge possible, that human life may be unhappy, and that my existence, if further prolonged, would become ineligible: but I thank providence, both for the good which I have already enjoyed, and for the power with which I am endowed of escaping the ill that threatens me.

DAVID HUME (1711–76), 'Of Suicide'

When would-be suicides in purpose fail—
Who could not find a morsel though they needed—
If Peter sends them for attempts to jail,
What would he do to them if they succeeded?

HOOD, 'Epigram'

Suicide, even publicly committed, remains the most private and impenetrable of human acts. The historian is unable to interrogate those who went to their chosen death silently, leaving no literature, no testament, indeed, in most cases, leaving scarcely anything at all save a poor bundle of clothing, small change, and broken objects, useful or useless, keys that had offered little solace, opening either on to misery and wretchedness, or on to promiscuity no longer tolerable (keys that at least would see further service), the staccato *états-civils* contained in the records drawn up *chez Daude*, telling of freckles, warts, spots, dimples, a shapely neck, a cleft chin, shape of nose, colour of hair—features that must, at one time, have been admired, or at least noticed, commented upon, loved, even caressed—and nothing at all about the once-living person; and, somewhere or other, nearly always not very far away and not at all difficult to discover, a room, or a corner of a room, with nothing in it that would give the slightest hint of individuality. Most had travelled light throughout their brief, truncated lives, taking with them only despair and hopelessness, and leaving behind them a rippling pool of regrets and temporary sadness, soon wiped away by time and forgetfulness, as if they had never existed, or perhaps lingering on a little more in the form of a guilty sense of relief, as the result of the removal of the only other witness, apart from conscience, to something shameful and tawdry, to cowardice, meanness, lack of imagination, a grinding selfishness, all likewise soon smoothed away, to restore the even surface of self-esteem, habit, and the daily effort spent to obtain the satisfaction of the basic needs.

It was no great achievement in life thus to have killed oneself; yet the *suicidés* and the *suicidées*, though they leave us nothing save the inexorable fact of their gesture and the summary decencies of their tattered and darned clothing, are as much mute chroniclers of their times, in the dreadfully repetitive record of their failures, as those who kept diaries, who wrote letters, and who had sufficient self-importance to feel that their *faits et gestes* were worth handing down to posterity. Failure is much commoner than success, at any period, though it has seldom been accorded even a small corner in the work of historians; it is also more endearing, and much more human. *No* death can ever be dismissed as banal, even if it cannot aspire to the proud luxury of a tombstone—a bold claim on the future—and death at one's own hand, a pitiable appeal for attention, an appeal quite unheard, cries out in anguish for ever.

RICHARD COBB, *Death in Paris, 1795–1801*, 1978

[*In 1732 Richard Smith, a bookbinder who had fallen into debt, and his wife decided to commit suicide. They killed their two-year-old daughter and hanged themselves, leaving the following letter, along with a note asking their landlord to look after their cat and dog.*]

These actions, considered in all their circumstances, being somewhat uncommon, it may not be improper to give some account of the cause; and that it was inveterate hatred we conceived against poverty and rags, evils that through a train of unlucky accidents were become inevitable. For we appeal to all that ever knew us, whether we were idle or extravagant, whether or no we have not taken as much pains to get our living as our neighbours, although not attended with the same success. We apprehend the taking of our child's life away to be a circumstance for which we shall be generally condemned; but for our own parts we are perfectly easy on that head. We are satisfied it is less cruelty to take the child with us, even supposing a state of annihilation as some dream of, than to leave her friendless in the world, exposed to ignorance and misery. Now in order to obviate some censures which may proceed either from ignorance or malice, we think it proper to inform the world, that we firmly believe the existence of an Almighty God; that this belief of ours is not an implicit faith, but deduced from the nature and reason of things. We believe the existence of an Almighty Being from the consideration of his wonderful works, from those innumerable celestial and glorious bodies, and from their wonderful order and harmony. We have also spent some time in viewing those wonders which are to be seen in the minute part of the world, and that with great pleasure and satisfaction. From all which particulars we are satisfied that such amazing things could not possibly be without a first mover—without the existence of an Almighty Being. And as we know the wonderful God to be Almighty, so we cannot help believing that he is also good—not implacable, not like such wretches as men are, not taking delight in the misery of his creatures; for which reason we resign up our breath to him without any terrible apprehensions, submitting ourselves to those ways which in his goodness he shall please to appoint after death. We also believe in the existence of unbodied natures, and think we have reason for that belief, although we do not pretend to know their way of subsisting. We are not ignorant of those laws made *in terrorem*, but leave the disposal of our bodies to the wisdom of the coroner and his jury, the thing being indifferent to us where our bodies are laid. From hence it will appear how little anxious we are about 'hic jacet' . . .

(Signed) RICHARD SMITH
BRIDGET SMITH

''Tis because of us children, too, isn't it, that you can't get a good lodging?'

'Well—people do object to children sometimes.'

'Then if children make so much trouble, why do people have 'em?'

'O—because it is a law of nature.'

'But we don't ask to be born?'

'No indeed.' . . .

'I think that whenever children be born that are not wanted they should be killed directly, before their souls come to 'em, and not allowed to grow big and walk about!'

Sue did not reply. She was doubtfully pondering how to treat this too reflective child.

She at last concluded that, so far as circumstances permitted, she would be honest and candid with one who entered into her difficulties like an aged friend.

'There is going to be another in our family soon,' she hesitatingly remarked.

'How?'

'There is going to be another baby.'

'What!' The boy jumped up wildly. 'O God, mother, you've never a-sent for another; and such trouble with what you've got!' . . .

He got up, and went away into the closet adjoining her room, in which a bed had been spread on the floor. There she heard him say: 'If we children was gone there'd be no trouble at all!'

'Don't think that, dear,' she cried, rather peremptorily. 'But go to sleep!'

Jude stood bending over the kettle, with his watch in his hand, timing the eggs, so that his back was turned to the little inner chamber where the children lay. A shriek from Sue suddenly caused him to start round. He saw that the door of the room, or rather closet—which had seemed to go heavily upon its hinges as she pushed it back—was open, and that Sue had sunk to the floor just within it. Hastening forward to pick her up he turned his eyes to the little bed spread on the boards; no children were there. He looked in bewilderment round the room. At the back of the door were fixed two hooks for hanging garments, and from these the forms of the two youngest children were suspended, by a piece of box-cord round each of their necks, while from a nail a few yards off the body of little Jude was hanging in a similar manner. An overturned chair was near the elder boy, and his glazed eyes were slanted into the room; but those of the girl and the baby were closed.

. . . a piece of paper was found upon the floor, on which was written, in the boy's hand, with the bit of lead pencil that he carried:

'Done because we are too menny.'

THOMAS HARDY (1840–1928), *Jude the Obscure*

On that night [in 1816] Fanny [Imlay], having arrived at the Mackworth Arms Inn, Swansea, by the Cambrian coach from Bristol, retired to rest, telling the chambermaid that she was exceedingly fatigued, and would herself take care of the candle. When she did not appear next morning they forced her chamber door, and found her lying dead; her long brown hair about her face; a bottle of laudanum upon the table, and a note which ran thus: 'I have long determined that the best thing I could do was to put an end to the existence of a being whose birth was unfortunate, and whose life has only been a series of pain to those persons who have hurt their health in endeavouring to promote her welfare. Perhaps to hear of my death will give you pain, but you will soon have the blessing of forgetting that such a creature ever existed as . . .' She had with her the little Genevan watch, a gift of travel from Mary and Shelley; and in her purse were a few shillings.

EDWARD DOWDEN, *The Life of Percy Bysshe Shelley*, 1896

My second sister, Carla, . . . had chosen a stage career, well equipped for it by her beauty but scarcely by any deeply rooted original gift. As a small child she had already been near death; a frightful complication of convulsions, whooping-cough, and inflammation of the lungs had made the doctors despair of her growing up. Her existence continued a frail and precarious one. A proud, disdainful nature, unconventional but refined, she loved literature, art, the manifestations of mind; and the crude unkindly time drove her into an unhappy bohemian existence. A taste for the macabre made her as a girl adorn her room with a death's head, to which she gave a scurrilous name; yet—the two things go very well together—she was as childishly laughter-loving as the rest of us . . . Disappointed in her professional aspirations, she remained the object of desire. Apparently she tried to find a way back into the bourgeois sphere, and her hopes centred about a marriage with the young son of an Alsatian industrialist who was in love with her. But she had before this given herself to another man, a doctor by profession, who used his power over her for his own gratification. The young fiancé found himself deceived and called her to account. Then she took her cyanide, enough to kill a whole company of soldiers.

. . . the betrothed had made his appearance. Coming from an interview with him, the unhappy creature hurried past her mother with a smile, locked herself into her room, and the last that was heard from her was the sound of the gargling with which she tried to cool the burning of her corroded throat. She had time, after that, to lie down on the couch. Dark spots on the hands and face showed that death by suffocation—after a brief delay—must have ensued very suddenly. A note in French was found: '*Je t'aime.* Une fois *je t'ai trompé, mais je t'aime.*'

THOMAS MANN, *A Sketch of My Life*, tr. H. T. Lowe-Porter

Dearest,

I feel certain that I am going mad again: I feel we cant go through another of those terrible times. And I shant recover this time. I begin to hear voices, and cant concentrate. So I am doing what seems the best thing to do. You have given me the greatest possible happiness. You have been in every way all that anyone could be. I dont think two people could have been happier till this terrible disease came. I cant fight it any longer, I know that I am spoiling your life, that without me you could work. And you will I know. You see I cant even write this properly. I cant read. What I want to say is that I owe all the happiness of my life to you. You have been entirely patient with me and incredibly good. I want to say that—everybody knows it. If anybody could have saved me it would have been you. Everything has gone from me but the certainty of your goodness. I cant go on spoiling your life any longer.

I dont think two people could have been happier than we have been.

V.

VIRGINIA WOOLF to her husband, Leonard, ?18 March 1941

Now, like the gods, he is invulnerable.

Nothing on earth can hurt him—neither a woman's rejection, nor his lung disease, nor the anxieties of verse, nor that white thing, the moon, which he need fix in words no more.

He walks slowly under the lindens. He looks at the balustrades and doors, but not to remember them.

He knows how many nights and days he has left.

His will has imposed a set discipline on him. So as to make the future as irrevocable as the past, he will carry out certain actions, cross determined street corners, touch a tree or a grille.

He does all this so that the act he desires and fears will be no more than the final term of a series.

He walks along Street 49. He thinks he will never make his way inside this or that entrance.

Without rousing suspicions, he has already said goodbye to many friends.

He thinks about what he will never know—whether the day after will be rainy.

He meets an acquaintance and tells him a joke. He knows that for a time this episode will furnish an anecdote.

Now, like the dead, he is invulnerable.

At the appointed hour, he will climb some marble steps. (This will be remembered by others.)

He will go down to the lavatory. There, on the chess-board-patterned floor tiles, water will wash the blood away quite soon. The mirror awaits him.

He will smooth back his hair, adjust his tie (as fits a young poet, he was always a bit of a dandy), and try to imagine that the other man—the one in the mirror—performs the actions and that he, the double, repeats them. His hand will not falter at the end. Obediently, magically, he will have pressed the weapon to his head.

It was in this way, I suppose, that things happened.

JORGE LUIS BORGES (b. 1899), 'May 20, 1928: On the Death of Francisco López Merino', tr. Norman Thomas di Giovanni

Alone, he came to his decision,
The sore tears stiffening his cheeks
As headlamps flicked the ceiling with white dusters
And darkness roared downhill with nervous brakes.
Below, the murmuring and laughter,
The baritone, tobacco-smelling jokes;
And then his misery and anger
Suddenly became articulate:
'I wish that I was dead. Oh, they'll be sorry then.
I hate them and I'll kill myself tomorrow.
I want to die. I hate them, hate them. Hate.'

And kill himself in fact he did,
But not next day as he'd decided.
The deed itself, for thirty years deferred,
Occurred one wintry night when he was loaded.
Belching with scotch and misery
He turned the gas tap on and placed his head
Gently, like a pudding, in the oven.
'I want to die. I'll hurt them yet,' he said,
And once again: 'I hate them, hate them. Hate.'
The lampless darkness roared inside his head,
Then sighed into a silence in which played
The grown-up voices, still up late,
Indifferent to his rage as to his fate.

VERNON SCANNELL (b. 1922), 'Felo de Se'

I have heard of a post-war writer who, after having finished his first book, committed suicide to attract attention to his work. Attention was in fact attracted, but the book was judged no good.

CAMUS, op. cit.

Numbers, however, will account for a great proportion of unbalanced and suffering humanity. One man will rove the streets seeking motor-cars with numbers that are divisible by seven. Well-known, alas, is the case of the poor German who was very fond of three and who made each aspect of his life a thing of triads. He went home one evening and drank three cups of tea with three lumps of sugar in each cup, cut his jugular with a razor three times and scrawled with a dying hand on a picture of his wife goodbye, goodbye, goodbye.

FLANN O'BRIEN (1911–66), *At Swim-Two-Birds*

Amongst these stylists there should be a place kept for Heliogabalus, unsuccessful as an emperor but unrivalled as an eccentric. The man who combed Arabia that the phoenix might grace his table, also prepared the instruments of his own death with meticulous forethought. Syrian priests had told him that he would end his life by suicide; and for this purpose he provided himself with a golden sword, poison enclosed in a priceless ring, and a rope of imperial purple and gold. Finally, in case none of these methods should suit his mood, he ordered a pavement of jewels to be laid beneath one of his towers, considering that only precious stones could decently receive an imperial precipitate. His forethought was unfortunately wasted, as his own guards murdered him.

Even the young Polish lady, who was unhappily in love, and over a period of five months swallowed four spoons, three knives, nineteen coins, twenty nails, seven window-bolts, a brass cross, one hundred and one pins, a stone, three pieces of glass and two beads from her rosary, only appals by her perseverance.

HENRY ROMILLY FEDDEN, *Suicide: A Social and Historical Study*, 1938

You, who can't do anything, think you can bring off something like that? How can you even dare to think about it? If you were capable of it, you certainly wouldn't be in need of it.

FRANZ KAFKA (1883–1924), of himself, in a letter, tr. Ronald Hayman

When a woman is in despair at the recent death of her lover on active service, and is obviously thinking of following him to the grave, one must first of all decide whether or not this would be a proper outcome. If the answer is negative, one must attack through that immemorial habit of humanity, her *instinct of self-preservation*. If the woman has an enemy, one can persuade her that this enemy has obtained a royal warrant for her summary imprisonment. If this threat does not increase her desire for death, she may begin to think of going into hiding to avoid incarceration. She should be in hiding for three weeks, fleeing from one refuge to another; she should be arrested, and escape three days later. Then under an assumed name she should be helped to find sanctuary in some very remote town as different as possible from the one where she was in despair. But who would wish to take up the cause of consoling so unhappy a being and one so unrewarding in friendship?

STENDHAL (1783–1842), *Love*, tr. Gilbert and Suzanne Sale

A practically infallible way of preserving yourself against the desire for self-destruction is always to have something to do. Creech, the commentator on Lucretius, noted on his manuscript: 'N.B. Must hang myself when I have finished.' He kept his word, that he might have the pleasure of ending like his author. Had he taken on a commentary upon Ovid, he would have lived longer.

VOLTAIRE, *Dictionnaire Philosophique*

Suicide is very contagious . . . There is the well-known story of the fifteen patients who hung themselves in swift succession in 1772 from the same hook in a dark passage of the hospital. Once the hook was removed there was an end of the epidemic. Likewise, at the camp of Boulogne, a soldier blew out his brains in a sentry-box; in a few days others imitated him in the same place; but as soon as this was burned, the contagion stopped. All these facts show the overpowering influence of obsession, because they cease with the disappearance of the material object which evoked the idea.

If there are countries which accumulate suicides and homicides, it is never in the same proportions; the two manifestations never reach their maximum intensity at the same point. It is even a general rule that where homicide is very common it confers a sort of immunity against suicide.

EMILE DURKHEIM (1858–1917), *Suicide: A Study in Sociology*, tr. John A. Spaulding and George Simpson

Modern drugs and domestic gas . . . have not only made suicide more or less painless, they have also made it seem magical. A man who takes a knife and slices deliberately across his throat is murdering himself. But when someone lies down in front of an unlit gas-fire or swallows sleeping pills, he seems not so much to be dying as merely seeking oblivion for a while. Dostoevsky's Kirilov said that there are only two reasons why we do not all kill ourselves: pain and the fear of the next world. We seem, more or less, to have got rid of both. In suicide, as in most other areas of activity, there has been a technological break-through which has made a cheap and relatively painless death democratically available to everyone. Perhaps this is why the subject now seems so central and so demanding, why even governments spend a little money on finding its causes and possible means of prevention. We already have a suicidology; all we mercifully lack, for the moment, is a thorough-going philosophical rationale of the act itself. No doubt it will come. But perhaps that is only as it should be in a period in which global suicide by nuclear warfare is a permanent possibility.

A. ALVAREZ (b. 1929), *The Savage God*

Sir, Thank you for your excellent leading article, 'The road to dusty death' (October 18). Ten years ago, while clinically depressed, I attempted suicide several times, sincerely believing that death was the 'only satisfactory release'. I thank God that I did not have access then to any guides to supposed self-deliverance.

I am now 33 years of age, have been happily married for seven years, have recently completed a book-keeping training course and gained employment in this field, am an active campaigner for human rights, and have an unshakeable religious faith.

Please, EXIT, give other people a chance to have a new life in *this* world.

Yours sincerely,
JEAN M. HASLAM

Letter to *The Times*, 24 October 1980

It is a decision that I shall not take, at least not yet, for the reason that I have got myself annuities from two sovereigns and I should be inconsolable if my death enriched two crowned heads.

VOLTAIRE, writing to Mme du Deffand, *Lettres*

How many people have wanted to kill themselves, and have been content with tearing up their photograph!

JULES RENARD (1864–1910), *Journal*

To brace and fortify the child who already is turning with fear and repugnance from the life he is born into, it is necessary to say: Things may easily become more than I choose to bear. That is a very healthy and a very positive attitude . . . that 'choose' is a grand old burn-your-boats phrase that will put beef into the little one, and you see if it doesn't bring him to a ripe old age. If he doesn't in the end go off natural I shall be surprised. Well look here, I am not paid anything for this statement, but look here, here I am. See what it's done for me. I'm twice the girl I was that lay crying and waiting for death to come at that convalescent home. No, when I sat up and said: Death has got to come if I call him, I never called him and never have.

So teach your little ones to look on Death as Thanatos-Hades the great Lord of the Dead, that must, great prince though he be, come to their calling. And on the shadowy wings of this dark prince let them be borne upwards from the mire of makeshift and fearful compromise.

STEVIE SMITH (1902–71), *Novel on Yellow Paper*

The thought of suicide is a great consolation: by means of it one gets successfully through many a bad night.

NIETZSCHE, *Beyond Good and Evil*

Of their goods and bodies we can dispose; but what shall become of their souls, God alone can tell; his mercy may come *inter pontem et fontem, inter gladium et jugulum*, betwixt the bridge and the brook, the knife and the throat. *Quod cuiquam contigit, cuivis potest*: who knows how he may be tempted? It is his case; it may be thine.

ROBERT BURTON (1577–1640), *The Anatomy of Melancholy*

Mourning

DEPLORING the modern taboo on mourning, psychologists and others have dwelt at some length on the human need to find an outlet for grief. Thus Geoffrey Gorer has called on society to provide secular 'ritual support' for the bereaved and their families and friends which would encourage and shape the expression of sorrow 'without embarrassment or reticence'.

'Ritual support', whether secular or religious and whatever the context, can soon decline into empty ritual. Such moving ceremonies as the playing of the last post depend for their power on their infrequency and specialness. Are we to have Requiem Masses on cassette? Diploma'd keeners and professional comforters? At all events, I feel little confidence either in the proposals (which are scant and largely unspecified) or in the assumptions behind them. I remember how, in youth, many of us reacted against the conventional paraphernalia of death and mourning, more prominent then than now—grandiose funerals among the well-to-do, black armbands and closed blinds among the rest, 'the trappings and the suits of woe', the performances touched on here by Henry Mayhew and Thomas Hardy—and we reckoned to find more sincerity in private and unadvertised sorrow.

A Chinese funeral which I attended more recently, in Singapore, did little to persuade me of the superiority of ritual. It was attended with loud wailing on the part of persons normally modest and dignified and the (dexterously not-quite) tearing of hair and rending of clothes. When the coffin had been lowered into the grave, the widow—restrained by her daughter with seeming difficulty—made as if to cast herself after it. No doubt it was only the few foreigners there who feared for a moment that she would really do so. This behaviour seemed all the more incongruous in that the dead man was a Chinese scholar of the old school, one who would have read Chuang Tzu on the death of his wife: 'To break in upon her rest with the noise of lamentation would but show that I knew nothing of nature's Sovereign Law.' But in this particular milieu, of diaspora Chinese, there could well be a disparity between the way women mourn men and the way men mourn women. And true, to submit to accepted practice can afford relief in a crisis. *Something* has to be done. The question is, how to create an accepted practice afresh.

It is easy to agree with what Solzhenitsyn says here on the subject of cemeteries, and what is said by others in the section following this one. Cemeteries should be places of common and relaxed resort, like our few remaining botanic gardens: 'the presence of the dead among the living will be a daily fact in any society which encourages its people to live.' The formalities customarily provided in crematorium chapels are thin and hurried, and, if of a secular nature, rely on music or poetry; we recall Matthew Arnold's prophecy, that 'most of what now passes with us for religion and philosophy will be replaced by poetry.' (Some of the best poetry, alas, lies in the burial service.) I dare say the psychologists are right in believing that we avert our eyes too readily for our own good; and society, or town planners, ought to be able to help here.

Whatever society's provisions, we still have to cope singly and for ourselves; the dead are *ours*, not society's. The point is made, albeit obliquely and with heavy irony, in another of Hardy's poems, 'In the Cemetery', describing mothers squabbling over whose child is buried in which grave. The caretaker confides to the reader that in fact the whole lot have been moved by order of the Council—but

> as well cry over a new-laid drain
> As anything else, to ease your pain!

For the rest, a contained, private mourning seems the more real and. in a society already violent and vulgar enough, the more decent, Probably there must always be tension between the guilty feeling of 'How could I forget thee?' and what the obstinately inconsolable mother in Robert Frost's 'Home Burial' calls with unfair scorn 'making the best of one's way back to life and living people'. Possibly a similar tension is present in our prefigurings of how in our turn we shall be mourned or otherwise. At moments we think we hope for a grand funeral in Westminster Abbey, nodding plumes, hats off all along the route, a damp handkerchief or two. More often we feel, like Joyce Grenfell, 'Weep if you must, but sing as well.' (The lawyer mentioned in the section on Graveyards and Funerals, who could not countenance a bequest towards a posthumous party, must have been exceptionally stuffy!)

Sorrow, Johnson wrote in *The Rambler*, is 'a kind of rust of the soul' and 'the putrefaction of stagnant life'. We would not truly wish such damage on those who are good enough to miss us.

The Bustle in a House
The Morning after Death
Is solemnest of industries
Enacted upon Earth—

The Sweeping up the Heart
And putting Love away
We shall not want to use again
Until Eternity.

EMILY DICKINSON, 'The Bustle in a House'

There is now a very general recognition that human beings do have sexual urges and that, if these are denied outlet, the result will be suffering, either psychological or physical or both. But there is no analogous secular recognition of the fact that human beings mourn in response to grief, and that, if mourning is denied outlet, the result will be suffering, either psychological or physical or both. At present death and mourning are treated with much the same prudery as sexual impulses were a century ago . . . It would seem correct to state that a society which denies mourning and gives no ritual support to mourners is thereby producing maladaptive and neurotic responses in a number of its citizens. And this further suggests the desirability of making social inventions which will provide secular mourning rituals for the bereaved, their kin and their friends and neighbours . . . Such rituals would have to take into account the need of the mourner for both companionship and privacy; the fact that it is (almost certainly) desirable for mourners to give expression to their grief without embarrassment or reticence; and the fact that for some weeks after bereavement a mourner is undergoing much the same physical changes as occur during and after a severe illness.

GEOFFREY GORER, *Death, Grief, and Mourning in Contemporary Britain*, 1965

Above all else we have grown to fear death and those who die.

If there is a death in a family we try to avoid writing or calling because we do not know what to say about death.

It is even considered shameful to mention a cemetery seriously. You would never say at work: 'Sorry, I can't come on Sunday, I've got to visit my relatives at the cemetery.' What is the point of bothering about people who are not going to invite you to a meal?

What an idea—moving a dead man from one town to another! No one would lend you a car for that. And nowadays, if you're a nonentity, you don't get a hearse and a funeral march—just a quick trip on a lorry.

Once people used to go to our cemeteries on Sundays and walk between the graves, singing beautiful hymns and spreading sweet-smelling incense. It set your heart at rest; it allayed the painful fears of inevitable death. It was almost as though the dead were smiling from under their grey mounds: 'It's all right . . . Don't be afraid.'

But nowadays, if a cemetery is kept up, there's a sign hanging there: 'Owners of graves! Keep this place tidy on penalty of a fine!' But more often they just roll them flat with bulldozers, to build sports grounds and parks.

Then there are those who died for their native land—it could still happen to you or me. There was a time when the church set aside a day of remembrance for those who fell on the battlefield. England does this on Poppy Day. All nations dedicate one day to remembering those who died for us all.

More men died for us Russians than for any other people, yet we have no such day. If you stop and think about the dead, who is to build the new world? In three wars we have lost so many husbands, sons and lovers—yet to think of them repels us. They're dead, buried under painted wooden posts—why should they interfere with our lives? For *we* will never die!

ALEXANDER SOLZHENITSYN (b. 1918), '*We* Will Never Die', tr. Michael Glenny

At this grief my heart was utterly darkened; and whatever I beheld was death. My native country was a torment to me, and my father's house a strange unhappiness; and whatever I had shared with him, wanting him, became a distracting torture. Mine eyes sought him everywhere, but he was not granted them; and I hated all places, for that they had not him; nor could they now tell me, 'he is coming', as when he was alive and absent. I became a great riddle to myself, and I asked my soul, *why she was so sad, and why she disquieted me sorely*: but she knew not what to answer me . . .

Thus was I wretched, and that wretched life I held dearer than my friend . . . for at once I loathed exceedingly to live, and feared to die. I suppose, the more I loved him, the more did I hate, and fear (as a most cruel enemy) death, which had bereaved me of him: and I imagined it would speedily make an end of all men, since it had power over him . . . I felt that my soul and his soul were 'one soul in two bodies': and therefore was my life a horror to me, because I would not live halved. And therefore perchance I feared to die, lest he whom I had much loved, should die wholly.

ST AUGUSTINE (354–430), on the loss of an early friend, *Confessions*, tr. E. B. Pusey

Dark house, by which once more I stand
Here in the unlovely street,
Doors, where my heart was used to beat
So quickly, waiting for a hand,

A hand that can be clasp'd no more—
Behold me, for I cannot sleep,
And like a guilty thing I creep
At earliest morning to the door.

He is not here; but far away
The noise of life begins again,
And ghastly thro' the drizzling rain
On the bald street breaks the blank day.

ALFRED LORD TENNYSON (1809–92), *In Memoriam A.H.H.*

That's the cuckoo, you say. I cannot hear it.
When last I heard it I cannot recall; but I know
Too well the year when first I failed to hear it—
It was drowned by my man groaning out to his sheep 'Ho! Ho!'

Ten times with an angry voice he shouted
'Ho! Ho!' but not in anger, for that was his way.
He died that Summer, and that is how I remember
The cuckoo calling, the children listening, and me saying, 'Nay.'

And now, as you said, 'There it is' I was hearing
Not the cuckoo at all, but my man's 'Ho! Ho!' instead.
And I think that even if I could lose my deafness
The cuckoo's note would be drowned by the voice of my dead.

EDWARD THOMAS, 'The Cuckoo'

They say one must keep your standards and your values of life alive. But how can I, when I only kept them for you? Everything was for you. I loved life just because you made it so perfect, and now there is no one left to make jokes with, or to talk about Racine and Molière and talk about plans and work and people.

I dreamt of you again last night. And when I woke up it was as if you had died afresh. Every day I find it *harder* to bear. For what point is there in life now? . . . I look at our favourites, I try and read them, but without you they give me no pleasure. I only remember the evenings when you read them to me aloud and then I cry. I feel as if we had collected all our wheat into a barn to make bread and beer for the rest of our lives and now our barn has been burnt down and we stand on a cold winter morning looking at the charred ruins. For this little room was the gleanings of our life together. All our happiness was over this fire and with these books. With Voltaire blessing us with upraised hand on the wall . . . It is impossible to think that I shall never sit with you again and hear your laugh. *That every day for the rest of my life you will be away.*

CARRINGTON, *Diaries*, 12/17 February 1932 (Lytton Strachey died in January 1932, and Carrington killed herself in March of that year)

Like the *yu'ub* wood bell tied to gelded camels that
    are running away,
Or like camels which are being separated from their young,
Or like people journeying while moving camp,
Or like a well which has broken its sides or a river
    which has overflowed its banks,
Or like an old woman whose only son was killed,
Or like the poor, dividing the scraps for their frugal meal,
Or like the bees entering their hive, or food crackling
    in the frying,
Yesterday my lamentations drove sleep from all the camps.
Have I been left bereft in my house and shelter?
Has the envy of others been miraculously fulfilled?
Have I been deprived of the fried meat and reserves for
    lean times which were so plentiful for me?
Have I today been taken from the chess-board?
Have I been borne on a saddle to a distant and desolate
    place?
Have I broken my shin, a bone which cannot be mended?

RAAGE UGAAS (19th century), a lament for his wife: *Somali Poetry: An Introduction*, B. W. Andrzejewski and I. M. Lewis

Why does the thin grey strand
Floating up from the forgotten
Cigarette between my fingers,
Why does it trouble me?

Ah, you will understand;
When I carried my mother downstairs,
A few times only, at the beginning
Of her soft-foot malady,

I should find, for a reprimand
To my gaiety, a few long grey hairs
On the breast of my coat; and one by one
I watched them float up the dark chimney.

LAWRENCE, 'Sorrow'

'You could sit there with the stains on your shoes
Of the fresh earth from your own baby's grave
And talk about your everyday concerns.
You had stood the spade up against the wall
Outside there in the entry, for I saw it.'

'I shall laugh the worst laugh I ever laughed.
I'm cursed. God, if I don't believe I'm cursed.'

'I can repeat the very words you were saying:
"Three foggy mornings and one rainy day
Will rot the best birch fence a man can build."
Think of it, talk like that at such a time!
What had how long it takes a birch to rot
To do with what was in the darkened parlour?
You *couldn't* care! The nearest friends can go
With anyone to death, comes so far short
They might as well not try to go at all.
No, from the time when one is sick to death,
One is alone, and he dies more alone.
Friends make pretence of following to the grave,
But before one is in it, their minds are turned
And making the best of their way back to life
And living people, and things they understand.
But the world's evil. I won't have grief so
If I can change it. Oh, I won't, I won't!' . . .

ROBERT FROST (1874–1963), from 'Home Burial'

The world of dew is
A world of dew, yet even
So, yet even so . . .

ISSA (1763–1827), composed shortly after the death of his only child

The shock was wearing off, but the ache was growing. I found myself setting the table for four. Something would have to be done about his bedroom: the closed door was almost as painful as the open. I found an exercise book with a piece of writing dated a month previously, in his small, untidy writing: *The Best Holiday Of My Life*. A visit to London: a jousting-tournament at the Tower, Regent's Park Zoo, the dinosaurs at the Natural History Museum. I put it at the back of a drawer, way back.

CHRISTOPHER LEACH, *Letter to a Younger Son*, 1981

All his beauty, wit and grace
Lie forever in one place.
He who sang and sprang and moved
Now, in death, is only loved.

ALICE THOMAS ELLIS, 'To Joshua', dedication prefixed to *The Birds of the Air*, 1980

Surprised by joy—impatient as the Wind
I turned to share the transport—Oh! with whom
But Thee, deep buried in the silent tomb,
That spot which no vicissitude can find?
Love, faithful love, recalled thee to my mind—
But how could I forget thee? Through what power,
Even for the least division of an hour,
Have I been so beguiled as to be blind
To my most grievous loss!—That thought's return
Was the worst pang that sorrow ever bore,
Save one, one only, when I stood forlorn,
Knowing my heart's best treasure was no more;
That neither present time, nor years unborn
Could to my sight that heavenly face restore.

WILLIAM WORDSWORTH (1770–1850), 'Surprised by joy—impatient as the Wind': 'suggested by my daughter Catharine long after her death'

The sun is soon to rise as bright
As if the night had brought no sorrow.
That grief belonged to me alone,
The sun shines on a common morrow.

You must not shut the night inside you,
But endlessly in light the dark immerse.
A tiny lamp has gone out in my tent—
I bless the flame that warms the universe.

FRIEDRICH RÜCKERT (1788–1866), *Songs on the Death of Children*

No different, I said, from rat's or chicken's,
That ten-week protoplasmic blob. But you
Cried as if you knew all that was nonsense
And knew that I did, too.

Well, I had to say something. And there
Seemed so little anyone could say.
That life had been in women's wombs before
And gone away?

This was our life. And yet, when the dead
Are mourned a little, then become unreal,
How should the never born be long remembered?
So this in time will heal

Though now I cannot comfort. As I go
The doctor reassures: 'Straightforward case.
You'll find, of course, it leaves her rather low.'
Something is gone from your face.

DAVID SUTTON (b. 1944), 'Not to be Born'

They're all gone now, and there isn't anything more the sea can do to me . . . I'll have no call now to be up crying and praying when the wind breaks from the south, and you can hear the surf is in the east, and the surf is in the west, making a great stir with the two noises, and they hitting one on the other. I'll have no call now to be going down and getting Holy Water in the dark nights after Samhain, and I won't care what way the sea is when the other women will be keening . . .

They're all together this time, and the end is come. May the Almighty God have mercy on Bartley's soul, and on Michael's soul, and on the souls of Sheamus and Patch, and Stephen and Shawn; and may he have mercy on my soul, and on the soul of every one is left living in the world . . . Michael has a clean burial in the far north, by the grace of the Almighty God. Bartley will have a fine coffin out of the white boards, and a deep grave surely. What more can we want than that? No man at all can be living for ever, and we must be satisfied.

J. M. SYNGE (1871–1909), *Riders to the Sea* (on an island off the West of Ireland, a mother mourns the sixth and last of her sons to be taken by the sea)

On my father's memorial day
I went out to see his mates—
All those buried with him in one row,
His life's graduation class.

I already remember most of their names,
Like a parent collecting his little son
From school, all of his friends.

My father still loves me, and I
Love him always, so I don't weep.
But in order to do justice to this place
I have lit a weeping in my eyes
With the help of a nearby grave—
A child's. 'Our little Yossy, who was
Four when he died.'

YEHUDA AMICHAI (b. 1924), 'My Father's Memorial Day', tr. the author and Ted Hughes

In vain to me the smiling mornings shine,
  And redd'ning Phoebus lifts his golden fire;
The birds in vain their amorous descant join,
  Or cheerful fields resume their green attire:
These ears alas! for other notes repine,
  A different object do these eyes require.
My lonely anguish melts no heart but mine,
  And in my breast the imperfect joys expire.
Yet morning smiles the busy race to cheer,
  And new-born pleasure brings to happier men;

The fields to all their wonted tribute bear,
  To warm their little loves the birds complain.
I fruitless mourn to him that cannot hear,
  And weep the more, because I weep in vain.

GRAY, 'Sonnet on the Death of Mr Richard West'

That the world will never be quite—what a cliché—the same again
Is what we only learn by the event
When a friend dies out on us and is not there
To share the periphery of a remembered scent

Or leave his thumb-print on a shared ideal;
Yet it is not at floodlit moments we miss him most,
Not intervolution of wind-rinsed plumage of oatfield
Nor curragh dancing off a primeval coast

Nor the full strings of passion; it is in killing
Time where he could have livened it, such as the drop-by-drop
Of games like darts or chess, turning the faucet
On full at a threat to the queen or double top.

LOUIS MACNEICE (1907–63), 'Tam Cari Capitis'

He went inside the café where they used to sit together.
It was here, three months ago, that his friend told him:
'We're completely broke—so hard up, the two of us,
that we're stuck with the cheapest places.
I can't go around with you any more—it's no use hiding the fact.
I've got to tell you, somebody else is after me.'
The 'somebody else' had promised him two suits, some silk
  handkerchiefs.
He himself, to get his friend back,
went through hell rounding up twenty pounds.
His friend came back to him for the twenty pounds—
but along with that, for their old intimacy,
their old love, for the deep feeling between them.
The 'somebody else' was a liar, a real bum:
he'd ordered only one suit for his friend,
and that under pressure, after much begging.

But now he doesn't want the suits any longer,
he doesn't want the silk handkerchiefs at all,
or twenty pounds, or twenty piastres even.

Sunday they buried him, at ten in the morning.
Sunday they buried him, almost a week ago.

He laid flowers on his cheap coffin,
lovely white flowers, very much in keeping
with his beauty, his twenty-two years.

When he went to the café that evening—
he happened to have some vital business there—
to that same café where they used to go together,
it was a knife in his heart,
that dead café where they used to go together.

C. P. Cavafy (1863–1933), 'Lovely White Flowers', tr. Edmund Keeley and Philip Sherrard

In thinking of all these virtues hold again, as it were, your son in your arms! He has now more leisure to devote to you, there is nothing now to call him away from you; never again will he cause you anxiety, never again any grief. The only sorrow you could possibly have had from a son so good is the sorrow you have had; all else is now exempt from the power of chance, and holds nought but pleasure if only you know how to enjoy your son, if only you come to understand what his truest value was. Only the image of your son—and a very imperfect likeness it was—has perished; he himself is eternal and has reached now a far better state, stripped of all outward encumbrances and left simply himself . . .

Do you therefore, Marcia, always act as if you knew that the eyes of your father and your son were set upon you—not such as you once knew them, but far loftier beings, dwelling in the highest heaven. Blush to have a low or common thought, and weep for those dear ones who have changed for the better! Throughout the free and boundless spaces of eternity they wander; no intervening seas block their course, no lofty mountains or pathless valleys or shallows of the shifting Syrtes; there every way is level, and, being swift and unencumbered, they easily are pervious to the matter of the stars and, in turn, are mingled with it.

Seneca, *Ad Marciam de Consolatione*, tr. J. W. Basore

When Chuang Tzu's wife died, Hui Tzu came to the house to join in the rites of mourning. To his surprise he found Chuang Tzu sitting with an inverted bowl on his knees, drumming upon it and singing a song. 'After all,' said Hui Tzu, 'she lived with you, brought up your children, grew old along with you. That you should not mourn for her is bad enough; but to let your friends find you drumming and singing—that is going too far!' 'You misjudge me,' said Chuang Tzu. 'When she died, I was in despair, as any man well might be. But soon, pondering on what had happened, I told myself that in death no strange new fate befalls us. In the beginning we lack not life only, but form. Not form only, but spirit. We are blended in the one great featureless indistinguishable mass. Then a time came when the mass evolved spirit, spirit evolved form, form evolved life. And now life in its turn has evolved death. For not nature only but man's being has its seasons, its sequence of spring and autumn, summer and winter. If some one is tired and has gone to lie down, we do not pursue him with shouting and bawling. She whom I have lost has lain down to sleep for a while in the Great Inner Room. To break in upon her rest with the noise of lamentation would but show that I knew nothing of nature's Sovereign Law. That is why I ceased to mourn.'

ARTHUR WALEY, *Three Ways of Thought in Ancient China*, 1939
(Chuang Tzu, Taoist philosopher, early 3rd century BC)

A slumber did my spirit seal;
  I had no human fears:
She seemed a thing that could not feel
  The touch of earthly years.

No motion has she now, no force;
  She neither hears nor sees;
Rolled round in earth's diurnal course,
  With rocks, and stones, and trees.

WORDSWORTH, 'A slumber did my spirit seal'

I know what it's like, he said, slapping him on the shoulder. I was like you, I was! After I lost my poor departed wife, I used to go off into the fields, to be on my own. I'd drop down at the foot of a tree, crying, calling on the good Lord, telling him off. I wanted to be like the moles, hung up on the branches, their bellies crawling with maggots—dead, I mean. And when I thought of the others, at that very moment, hugging their bonny little wives close to them, I'd strike the ground with my stick, hard. I was crazy, like, didn't even eat, the thought of going to the inn sickened me, you wouldn't believe it. Ah well, slowly but surely, one day chasing another, spring on top of winter, autumn on top of summer, it leaked away, drop by drop, little by little; it left, it went away—it sank down, I should say, because there's always something stays, at the bottom, so to speak . . . a weight there, on the chest! But it's the same for all of us, we mustn't let ourselves go, and want to die just because others are dead . . . You must pull yourself together, Monsieur Bovary—it will pass! Come and see us. My daughter thinks of you from time to time, you know, and you are forgetting her, so she says. It'll be spring soon; we'll get you to shoot a rabbit in the woods, to help take your mind off things.

GUSTAVE FLAUBERT (1821–80), *Madame Bovary*

What's that cart that nobody sees
grinding along the shore road?

Whose is the horse that pulls it, the white horse
that bares its yellow teeth to the wind?

They turn, unnoticed by anyone,
into the field of slanted stones.

My friends meet me. They lift me from the cart and,
the greetings over, we go smiling underground.

NORMAN MACCAIG (b. 1910), 'Every day'

. . . as I have discovered, passionate grief does not link us with the dead but cuts us off from them. This becomes clearer and clearer. It is just at those moments when I feel least sorrow—getting into my morning bath is one of them—that H. rushes upon my mind in her full reality, her otherness. Not, as in my worst moments, all foreshortened and patheticized and solemnized by miseries, but as she is in her own right. This is good and tonic.

C. S. LEWIS (1898–1963), *A Grief Observed*

It is well known that mourners often get the illness which led to the death of a close person. Habits and interests of the deceased may be taken over indiscriminately . . . A hitherto rather dull wife whose witty husband had died surprised herself and all around her by her newly acquired gift of repartee. She tried to explain this by saying alternately, 'I have to do it for him now' or, 'It isn't really me, he speaks out of me' (like a ventriloquist). This same woman, partner in a very good, loving marriage, told me that she had always been amused by her husband's patient peeling of the top of his boiled egg, while she used to cut it off. 'Now,' she said, 'I just cannot bring myself to cut the top off, I have to peel it off patiently.'

LILY PINCUS, *Death and the Family*, 1976

BOSWELL: 'But suppose now, Sir, that one of your intimate friends were apprehended for an offence for which he might be hanged.' JOHNSON: 'I should do what I could to bail him, and give him any other assistance; but if he were once fairly hanged, I should not suffer.' BOSWELL: 'Would you eat your dinner that day, Sir?' JOHNSON: 'Yes, Sir; and eat it as if he were eating it with me. Why, there's Baretti, who is to be tried for his life tomorrow, friends have risen up for him on every side; yet if he should be hanged, none of them will eat a slice of plum-pudding the less. Sir, that sympathetic feeling goes a very little way in depressing the mind.'

BOSWELL, *Life of Johnson*

The truth is, nobody suffered more from pungent sorrow at a friend's death than Johnson, though he would suffer no one else to complain of their losses in the same way; 'for (says he) we must either outlive our friends you know, or our friends must outlive us; and I see no man that would hesitate about the choice.'

HESTER LYNCH PIOZZI, *Anecdotes of the late Samuel Johnson*, 1786

'Such a Niobe you never saw.'

'Was she weeping?'

'Not actual tears. But her gown, and her cap, and her strings were weeping. Her voice wept, and her hair, and her nose, and her mouth. Don't you know that look of subdued mourning?'

ANTHONY TROLLOPE (1815–82), *The Prime Minister* (the Duchess of Omnium on the widowed Mrs Lopez)

The great Mrs Churchill was no more . . . Goldsmith tells us that when lovely woman stoops to folly, she has nothing to do but die; and when she stoops to be disagreeable, it is equally to be recommended as a clearer of ill fame. Mrs Churchill, after being disliked at least twenty-five years, was now spoken of with compassionate allowances.

JANE AUSTEN (1775–1817), *Emma*

Here shift the scene, to represent
How those I love my death lament.
Poor Pope will grieve a month; and Gay
A week; and Arbuthnot a day.

St John himself will scarce forbear
To bite his pen and drop a tear.
The rest will give a shrug, and cry,
'I'm sorry, but we all must die.'
Indifference clad in wisdom's guise,
All fortitude of mind supplies;
For how can stony bowels melt
In those who never pity felt?
When *we* are lashed, *they* kiss the rod,
Resigning to the will of God.

The fools, my juniors by a year,
Are tortured with suspense and fear;
Who wisely thought my age a screen,
When death approached, to stand between:
The screen removed, their hearts are trembling;
They mourn for me without dissembling.

My female friends, whose tender hearts
Have better learned to act their parts,
Receive the news in doleful dumps:
'The Dean is dead (*and what is trumps?*)'—
'Then Lord have mercy on his soul.
(*Ladies, I'll venture for the Vole.*)'—
'Six deans, they say, must bear the pall.
(*I wish I knew what King to call.*)
Madam, your husband will attend
The funeral of so good a friend?'—
'No, madam; 'tis a shocking sight,
And he's engaged tomorrow night!
My Lady Club would take it ill
If he should fail her at quadrille.
He loved the Dean. (*I led a heart.*)
But dearest friends, they say, must part.
His time was come, he ran his race;
We hope he's in a better place.' . . .

SWIFT, from 'Verses on the Death of Dr Swift'

Shopman.— . . . How deep would you choose to go, ma'am? Do you wish to be very poignant?

Lady.—Why, I suppose, crape and bombazine, unless they're gone out of fashion . . .

Shopman.— . . . We have a very extensive assortment, whether for family, court, or complimentary mourning; including the last novelties from the Continent.

Lady.—Yes; I should like to see them.

Shopman.—Certainly. Here is one, ma'am, just imported—a widow's silk—watered, as you perceive, to match the sentiment. It is called the 'Inconsolable', and is very much in vogue, in Paris, for matrimonial bereavements.

Squire.—Looks rather flimsy though. Not likely to last long; eh, sir? . . .

Shopman.— . . . several new fabrics have been introduced to meet the demand for fashionable tribulation.

Lady.—And all in the French style?

Shopman.—Certainly—of course, ma'am. They excel in the *funèbre*. Here, for instance, is an article for the deeply afflicted. A black crape, expressly adapted to the proposed style of mourning—makes up very sombre and interesting . . . Or, if you would prefer a velvet, ma'am—

Lady.—Is it proper, sir, to mourn in velvet?

Shopman.—O quite!—certainly. Just coming in. Now here is a very rich one—real Genoa—and a splendid black. We call it 'The Luxury of Woe'.

Lady.—Very expensive, of course?

Shopman.—Only eighteen shillings a yard, and a superb quality—in short, fit for the handsomest style of domestic calamity . . . The mourning of the poor people is very coarse—very; quite different from that of persons of quality—canvas to crape . . .

Lady.—To be sure it is! And as to the change of dress, sir; I suppose you have a great variety of half-mourning?

Shopman.—Oh! infinite—the largest stock in town. Full, and half, and quarter, and half-quarter, shaded off, if I may say so, like an India-ink drawing, from a grief *prononcé* to the slightest *nuance* of regret.

HENRY MAYHEW, ed., *The Shops and Companies of London*, 1865 (quoting from *Hoods Magazine*)

The widow's dress was uncomfortable. The streamers on the cap, because of their weight and roughness (they caught on the dress) made it difficult to turn the head. The dress was said also to be 'unhygienic', and exercise was 'impossible'. With the dress went a social ostracism;

for the first year no invitations could be accepted, and it was 'in the worst taste to be seen in places of public resort'. After the first year, the widow could, gradually, resume her place in society.

The etiquette for the mourning of a child for a parent was the same as that of a parent for a child; twelve months in all was required. For the first three paramatta, merino, and such dull cloths were worn, with a good deal of crape; the latter was usually arranged in two very deep tucks. For the next three months mourning silk was worn, with less crape, the crape ornamentally arranged in folds, plaits, or bouillonnés (our age, accustomed for many years to plain surfaces in dress, perhaps finds it difficult to realize how deprived the Victorian woman would feel robbed of the elaborate trimmings that helped to mark her status). The bonnet was trimmed with jet, and had a net veil with a deep crape hem: 'Linen collars and cuffs cannot be worn with crape; crêpe lisse frills are *de rigueur*. Sable or any coloured fur must be left off; sealskin is admissible, but it never looks well in really deep mourning.' After six months, crape could be abandoned for plain black, with jet ornaments and black gloves. This, after another two months, could be augmented with gold and black gloves. Then half mourning could be worn. Society was to be relinquished totally for two months.

JOHN MORLEY, *Death, Heaven and the Victorians*, 1971 (referring to the early 1880s)

'I stood at the back of the shop, my dear,
  But you did not perceive me.
Well, when they deliver what you were shown
  I shall know nothing of it, believe me!'

And he coughed and coughed as she paled and said,
  'O, I didn't see you come in there—
Why couldn't you speak?'—'Well, I didn't. I left
  That you should not notice I'd been there.

You were viewing some lovely things. "*Soon required
  For a widow, of latest fashion*";
And I knew 'twould upset you to meet the man
  Who had to be cold and ashen

And screwed in a box before they could dress you
  "*In the last new note in mourning,*"
As they defined it. So, not to distress you,
  I left you to your adorning.'

HARDY, 'At the Draper's'

Make bitter weeping, and make passionate wailing, and let thy mourning be according to his desert, for a day or two, lest thou be evil spoken of: and so be comforted for thy sorrow. For of sorrow cometh death, and sorrow of heart will bow down the strength . . . Give not thy heart unto sorrow: put it away, remembering the last end: forget it not, for there is no returning again: him thou shalt not profit, and thou wilt hurt thyself. Remember the sentence upon him; for so also shall thine be; yesterday for me, and today for thee.

Ecclesiasticus, 38

My answer to Saint Paul's question 'O death, where is thy sting?' is Saint Paul's own answer: 'The sting of death is sin.' The sin that I mean is the sin of selfishly failing to wish to survive the death of someone with whose life my own life is bound up. This is selfish because the sting of death is less sharp for the person who dies than it is for the bereaved survivor. This is, as I see it, the capital fact about the relation between living and dying. There are two parties to the suffering that death inflicts; and, in the apportionment of this suffering, the survivor takes the brunt.

ARNOLD TOYNBEE, *Man's Concern With Death*

If I should go before the rest of you
Break not a flower nor inscribe a stone,
Nor when I'm gone speak in a Sunday voice
But be the usual selves that I have known.
Weep if you must,
Parting is hell,
But life goes on,
So sing as well.

JOYCE GRENFELL (1910–79), *Joyce: by Herself and Her Friends*

When I am dead, my dearest,
Sing no sad songs for me;
Plant thou no roses at my head,
Nor shady cypress tree:
Be the green grass above me
With showers and dewdrops wet;
And if thou wilt, remember,
And if thou wilt, forget.

I shall not see the shadows,
  I shall not feel the rain;
I shall not hear the nightingale
  Sing on, as if in pain;
And dreaming through the twilight
  That doth not rise nor set,
Haply I may remember,
  And haply may forget.

CHRISTINA ROSSETTI, 'Song'

'The wind doth blow today, my love,
  And a few small drops of rain;
I never had but one truelove,
  In cold grave she was lain.

I'll do as much for my truelove
  As any young man may;
I'll sit and mourn all at her grave
  For a twelvemonth, and a day.'

The twelvemonth and a day being up,
  The dead began to speak,
'Oh who sits weeping on my grave,
  And will not let me sleep?'

''Tis I, my love, sits on your grave
  And will not let you sleep,
For I crave one kiss of your clay-cold lips
  And that is all I seek.'

'You crave one kiss of my clay-cold lips,
  But my breath smells earthy strong;
If you have one kiss of my clay-cold lips
  Your time will not be long:

'Tis down in yonder garden green,
  Love, where we used to walk,
The finest flower that ere was seen
  Is withered to a stalk.

The stalk is withered dry, my love,
  So will our hearts decay;
So make yourself content, my love,
  Till God calls you away.'

ANON., 'The Unquiet Grave'

Even beauty must die! That which subdues both gods and mortals
  Leaves the steely breast of the Stygian Zeus unmoved.
Once and once only did love soften the Lord of the Shadows,
  Then, on the very threshold, he sternly revoked his gift.
Aphrodite herself has no power to assuage the hurt
  Of her lover, his tender flesh ripped by the cruel boar.
Nor can the godlike hero be saved by his deathless mother
  At the Scaean gate when, falling, he achieves his fate:
She rises out of the sea with all the daughters of Nereus,
  And the mourning begins for her glorious son.
See! Where the gods are weeping, and the goddesses, all of them,
  Weeping that beauty passes, that perfection will die.
Yet to be a lament in the mouths of our loved ones is splendid,
  For what is common goes down into Orcus unsung.

FRIEDRICH VON SCHILLER (1759–1805), 'Lament'

Graveyards and Funerals

SYLVIA TOWNSEND WARNER paints for us an ideal picture of the kind of cemetery envisaged in the preface to the foregoing section. Death, exactly as Geoffrey Gorer and others would have it, 'is not out of fashion'. Montale, on the other hand, describes a commoner reality of our urban day: 'a few cypresses, second-rate tombs with artificial flowers . . .' While there is no attempt here to provide a history, however skeletonic, of burial customs and places, I have included a few ancient and far-flung instances, among them the life-loving tombs of Lawrence's Etruscans, the charming domesticity of the Mauretanian inscription, and the ostensibly less appetizing though equally pious feasting of the Melanesians as recorded by Malinowski.

Despite what so many have said about the democracy of the grave, 'the house appointed for all living', fashion in another sense of the word has been with us for a long time—with the worldly aspirations of the Renaissance in mind, Bosola remarks that men 'affect fashion in the grave'—and is with us still. It has been conjectured (though another theory favours a derivation from bedsteads) that the catch-phrase 'with knobs on' comes from an old popular song about how fine it would be to be 'blooming well dead' and in a coffin 'with knobs on'. 'Fashion' may take the form of contempt for the attitudes of other classes or races, as in Jilly Cooper's reference to the 'lower middle busybody' and Philip Freneau's comically patronizing reflections on the backward Indians who fail to understand that the dead are dead. It may manifest itself in the splendid but unfortunate funeral reported by Horace Walpole and the splendid funeral unfortunately immortalized by William McGonagall. Or, latterly, in large-scale financial investment in life everlasting, as conveyed in the *Times* dispatch from Los Angeles. Man is 'pompous in the grave', according to Sir Thomas Browne: but even more pompous in a refrigerator. Mr Mould was right after all: 'Do not let us say that gold is dross, when it can buy such things as these, Mrs Gamp.'

It is a relief to turn to Yuan Mei's refined notes on etiquette—the women can come and wail over his body just once, but monks and clappers are strictly forbidden—and a pleasure to end with the Greek vision of a dwelling-place in the stars.

Man is a Noble Animal, splendid in ashes, and pompous in the grave.

SIR THOMAS BROWNE (1605–82), *Urn Burial*

No people who turn their backs on death can be alive. The presence of the dead among the living will be a daily fact in any society which encourages its people to live. Huge cemeteries on the outskirts of cities, or in places no one ever visits, impersonal funeral rites, taboos which hide the fact of death from children, all conspire to keep the fact of death away from us, the living . . .

Never build massive cemeteries. Instead, allocate pieces of land throughout the community as grave sites—corners of parks, sections of paths, gardens, beside gateways—where memorials to people who have died can be ritually placed with inscriptions and mementoes which celebrate their life. Give each grave site an edge, a path, and a quiet corner where people can sit. By custom, this is hallowed land.

CHRISTOPHER ALEXANDER, SARA ISHIKAWA, MURRAY SILVERSTEIN *et al.*, *A Pattern Language: Towns, Buildings, Construction*, 1977

Still in the countryside among the lowly
Death is not out of fashion,
Still is the churchyard park and promenade
And a new-made grave a glory.
Still on Sunday afternoons, contentedly and slowly,
Come widows eased of their passion,
Whose children flitting from stone to headstone façade
Spell out accustomed names and the same story.

From mound to mound chirps grasshopper to grasshopper:
John dear husband of Mary,
Ada, relict, Lydia the only child,
Seem taking part in the chatter.
With boom and stumble, with cadence and patient cropper,
The organist practises the voluntary,
Swallows rehearsing their flight sit Indian-filed,
And under the blue sky nothing is the matter.

With spruce asters and September roses
Replenished are jampots and vases,
From the breasts of the dead the dead blossoms are swept
And tossed over into the meadow.
Women wander from grave to grave inspecting the posies,
And so tranquilly time passes
One might believe the scything greybeard slept
In the yew tree's shadow.

Here for those that mourn and are heavy-laden
Is pledge of Christ's entertainment;
Here is no Monday rising from warm bed,
No washing or baking or brewing,
No fret for stubborn son or flighty maiden,
No care for food or raiment;
No sweeping or dusting or polishing need the dead,
Nothing but flowers' renewing.

Here can the widow walk and the trembling mother
And hear with the organ blended
The swallows' auguring twitter of a brief flight
To a securer staying;
Can foretaste that heavenly park where toil and pother,
Labour and sorrow ended,
They shall stroll with husband and children in blameless white,
In sunlight, with music playing.

SYLVIA TOWNSEND WARNER (1893–1978), 'Graveyard in Norfolk'

To an impervious nothingness they're thinned,
For the red clay has swallowed the white kind;
Into the flowers that gift of life has passed.
Where are the dead?—their homely turns of speech,
The personal grace, the soul informing each?
Grubs thread their way where tears were once composed.

The bird-sharp cries of girls whom love is teasing,
The eyes, the teeth, the eyelids moistly closing,
The pretty breast that gambles with the flame,
The crimson blood shining when lips are yielded,
The last gift, and the fingers that would shield it—
All go to earth, go back into the game.

PAUL VALÉRY (1871–1945), from 'The Graveyard by the Sea', tr. C. Day Lewis

Row after row with strict impunity
The headstones yield their names to the element,
The wind whirrs without recollection;
In the riven troughs the splayed leaves
Pile up, of nature the casual sacrament
To the seasonal eternity of death;
Then driven by the fierce scrutiny
Of heaven to their election in the vast breath,
They sough the rumour of mortality . . .

ALLEN TATE (1899–1978), from 'Ode to the Confederate Dead'

If I could have lived another year
I could have finished my flying machine,
And become rich and famous.
Hence it is fitting the workman
Who tried to chisel a dove for me
Made it look more like a chicken.
For what is it all but being hatched,
And running about the yard,
To the day of the block?
Save that a man has an angel's brain,
And sees the axe from the first!

EDGAR LEE MASTERS (1868–1950), 'Franklin Jones'

In life I was the town drunkard;
When I died the priest denied me burial
In holy ground.
The which redounded to my good fortune.
For the Protestants bought this lot,
And buried my body here,
Close to the grave of the banker Nicholas,
And of his wife Priscilla.
Take note, ye prudent and pious souls,
Of the cross-currents in life
Which bring honour to the dead, who lived in shame.

As above, 'Chase Henry'

Upon Milton's grave there is supposed to have been no memorial; but in our time a monument has been erected in Westminster Abbey *To the Author of Paradise Lost*, by Mr Benson, who has in the inscription bestowed more words upon himself than upon Milton.

When the inscription for the monument of Philips, in which he was said to be *soli Miltono secundus*, was exhibited to Dr Sprat, then dean of Westminster, he refused to admit it; the name of Milton was, in his opinion, too detestable to be read on the wall of a building dedicated to devotion. Atterbury, who succeeded him, being author of the inscription, permitted its reception. 'And such has been the change of public opinion,' said Dr Gregory, from whom I heard this account, 'that I have seen erected in the church a statue of that man, whose name I once knew considered as a pollution of its walls.'

JOHNSON, *Life of Milton*

It isn't pleasant
to know that you are under the earth
even if the place is like an Island of the Dead
with a suggestion of Renaissance. It isn't
pleasant to think about it
but it's worse to see. A few cypresses,
second-rate tombs with artificial
flowers, and outside a small
parking lot for improbable
vehicles. But I know
that these dead lived a few steps from here,
you were an exception. I am
horrified to think that what's inside,
four bones and a few trinkets was
believed to be all of you
and perhaps it was, though it's terrible to say so.
Maybe you left in a hurry thinking
that he who makes the first move
gets the best place. But what place
and where? One continues to think
with a human head while entering the inhuman.

EUGENIO MONTALE (1896–1981), 'The Inhuman', tr. G. Singh

Duchess. Thou art very plain.
Bosola. My trade is to flatter the dead, not the living—I am a tomb-maker.
Duch. And thou comest to make my tomb?
Bos. Yes.
Duch. Let me be a little merry—of what stuff wilt thou make it?

Bos. Nay, resolve me first, of what fashion?
Duch. Why, do we grow fantastical in our death-bed? do we affect fashion in the grave?
Bos. Most ambitiously: princes' images on their tombs do not lie, as they were wont, seeming to pray up to heaven, but with their hands under their cheeks, as if they died of the tooth-ache; they are not carved with their eyes fixed upon the stars, but as their minds were wholly bent upon the world, the selfsame way they seem to turn their faces.

WEBSTER, *The Duchess of Malfi*

The caretaker hung his thumbs in the loops of his gold watch chain and spoke in a discreet tone to their vacant smiles.

— They tell the story, he said, that two drunks came out here one foggy evening to look for the grave of a friend of theirs. They asked for Mulcahy from the Coombe and were told where he was buried. After traipsing about in the fog they found the grave, sure enough. One of the drunks spelt out the name: Terence Mulcahy. The other drunk was blinking up at a statue of our Saviour the widow had got put up.

The caretaker blinked up at one of the sepulchres they passed. He resumed:

— And, after blinking up at the sacred figure, *Not a bloody bit like the man*, says he. *That's not Mulcahy*, says he, *whoever done it*.

Rewarded by smiles he fell back and spoke with Corny Kelleher . . .

— That's all done with a purpose, Martin Cunningham explained to Hynes.

— I know, Hynes said, I know that.

— To cheer a fellow up, Martin Cunningham said. It's pure goodheartedness.

JAMES JOYCE (1882–1941), *Ulysses*

In these customs is clearly expressed the fundamental attitude of mind of the surviving relative, friend or lover, the longing for all that remains of the dead person and the disgust and fear of the dreadful transformation wrought by death.

One extreme and interesting variety in which this double-edged attitude is expressed in a gruesome manner is sarco-cannibalism, a custom of partaking in piety of the flesh of the dead person. It is done with extreme repugnance and dread and usually followed by a violent vomiting fit. At the same time it is felt to be a supreme act of reverence, love, and devotion. In fact it is considered such a sacred duty that among the Melanesians of New Guinea, where I have studied and witnessed it, it is still performed in secret, although severely penalized by the white Government.

In all such rites, there is a desire to maintain the tie and the parallel tendency to break the bond. Thus the funerary rites are considered as unclean and soiling, the contact with the corpse as defiling and dangerous, and the performers have to wash, cleanse their body, remove all traces of contact, and perform ritual lustrations. Yet the mortuary ritual compels man to overcome the repugnance, to conquer his fears, to make piety and attachment triumphant, and with it the belief in a future life, in the survival of the spirit.

BRONISŁAW MALINOWSKI (1884–1942), *Magic, Science and Religion*

Excellent ritual of oils, of anointing,
office of priests;
everything was paid before these dead put on
the armless dress of their sarcophagi,
lying down in Phoenicia,

pillowing their heavy sculptured heads, their broad
foreheads like rides of sand, the rock of the chin,
the mouth, the simple map of the face, the carved
hair in full sail.

Surrender of sunlight and market and the white
loops of the coast,
was simply a journey, a bargain rigid as stone:
though youth took passage.

Stretched by the salt and echoing roads to the West
twenty-six bargain-makers of Phoenicia;
twenty-six dead with wide eyes,
confident of harbour.

BERNARD SPENCER (1909–63), 'Sarcophagi'

They are surprisingly big and handsome, these homes of the dead. Cut out of the living rock, they are just like houses. The roof has a beam cut to imitate the roof-beam of the house. It is a house, a home . . .

The tombs seem so easy and friendly, cut out of rock underground.

One does not feel oppressed, descending into them. It must be partly owing to the peculiar charm of natural proportion which is in all Etruscan things of the unspoilt, unromanized centuries. There is a simplicity, combined with a most peculiar, free-breasted naturalness and spontaneity, in the shapes and movements of the underworld walls and spaces, that at once reassures the spirit. The Greeks sought to make an impression, and Gothic still more seeks to impress the mind. The Etruscans, no. The things they did, in their easy centuries, are as natural and as easy as breathing. They leave the breast breathing freely and pleasantly, with a certain fullness of life. Even the tombs. And that is the true Etruscan quality: ease, naturalness, and an abundance of life, no need to force the mind or the soul in any direction.

And death, to the Etruscans, was a pleasant continuance of life, with jewels and wine and flutes playing for the dance. It was neither an ecstasy of bliss, a heaven, nor a purgatory of torment. It was just a natural continuance of the fullness of life. Everything was in terms of life, of living.

LAWRENCE, *Etruscan Places*

I would rather sleep in the southern corner of a little country churchyard, than in the tomb of the Capulets. I should like, however, that my dust should mingle with kindred dust.

EDMUND BURKE (1729–97), letter to Matthew Smith

We all set out the furnishings suited to a worthy grave,
And on the altar that marks the tomb of our mother,
  Secundula,
It pleased us to place a stone table-top,
Where we could sit around, bringing to memory her many
  good deeds,
As the food and the drinking cups were set out, and
  cushions piled around,
So that the bitter wound that gnawed our hearts might be
  healed.
And in this way we passed the evening hours in pleasant
  talk,
And in the praise of our good mother.
The old lady sleeps: she who fed us all
Lies silent now, and sober as ever.

Inscription from Mauretania (late third century)

This very day, a little while ago, you lived
But now you are neither man nor woman,
Breathless you are, for the Navahos killed you!
Then remember us not, for here and now
We bring you your food. Then take and keep
Your earth-walled place: once! twice!
Three times! four times! Then leave us now!

Tewa (Pueblo Indian) song, tr. H. J. Spinden

In spite of all the learned have said,
I still my old opinion keep;
The *posture*, that *we* give the dead,
Points out the soul's eternal sleep.

Not so the ancients of these lands—
The Indian, when from life released,
Again is seated with his friends,
And shares again the joyous feast.

His imaged birds, and painted bowl,
And venison, for a journey dressed,
Bespeak the nature of the soul,
ACTIVITY, that knows no rest.

His bow, for action ready bent,
And arrows, with a head of stone,
Can only mean that life is spent,
And not the old ideas gone.

Thou, stranger, that shalt come this way,
No fraud upon the dead commit—
Observe the swelling turf, and say
They do not *lie*, but here they *sit* . . .

PHILIP FRENEAU (1752–1832), from 'The Indian Burying Ground'

At the hour shaped for him Scyld departed,
the many-strengthed moved into his Master's keeping.

They carried him out to the current sea,
his sworn arms-fellows, as he himself had asked
while he wielded by his words, Ward of the Scyldings,
beloved folk-founder; long had he ruled.

A boat with a ringed neck rode in the haven,
icy, out-eager, the aetheling's vessel,
and there they laid out their lord and master,
dealer of wound gold, in the waist of the ship,
in majesty by the mast.
A mound of treasures
from far countries was fetched aboard her,
and it is said that no boat was ever more bravely fitted out
with the weapons of a warrior, war accoutrement,
bills and byrnies; on his breast were set
treasures and trappings to travel with him
on his far faring into the flood's sway.

This hoard was not less great than the gifts he had
from those who sent him, on the sill of life,
over seas, alone, a small child.

High over head they hoisted and fixed
a gold *signum*; gave him to the flood,
let the seas take him, with sour hearts
and mourning moods. Men have not the knowledge
to say with any truth—however tall beneath the heavens,
however much listened to—who unloaded that boat.

Beowulf (*c.*720), tr. Michael Alexander (the funeral of Scyld Shefing)

Three white wands had been stuck in the sand to mark the Poet's grave, but as they were at some distance from each other, we had to cut a trench thirty yards in length, in the line of the sticks, to ascertain the exact spot, and it was nearly an hour before we came upon the grave . . .

The lonely and grand scenery that surrounded us so exactly harmonized with Shelley's genius, that I could imagine his spirit soaring over us. The sea, with the islands of Gorgona, Capraja, and Elba, was before us; old battlemented watch-towers stretched along the coast, backed by the marble-crested Apennines glistening in the sun, picturesque from their diversified outlines, and not a human dwelling was in sight. As I thought of the delight Shelley felt in such scenes of loneliness and grandeur whilst living, I felt we were no better than a herd of wolves or a pack of wild dogs, in tearing out his battered and naked body from the pure yellow sand that lay so lightly over it, to drag him back to the light of day; but the dead have no voice, nor had I power to check the sacrilege—the work went on silently in the deep and unresisting sand, not a word was spoken, for the Italians have a touch of sentiment, and their feelings are easily excited into sympathy. Even Byron was silent and thoughtful. We were startled and drawn together by a dull hollow sound that followed the blow of a mattock; the iron had struck a skull, and the body was soon uncovered. Lime had been strewn on it; this, or decomposition, had the effect of staining it of a dark and ghastly indigo colour. Byron asked me to preserve the skull for him; but remembering that he had formerly used one as a drinking-cup, I was determined Shelley's should not be so profaned . . . After the fire was well kindled . . . more wine was poured over Shelley's dead body than he had consumed during his life. This with the oil and salt made the yellow flames glisten and quiver. The heat from the sun and fire was so intense that the atmosphere was tremulous and wavy. The corpse fell open and the heart was laid bare. The frontal bone of the skull, where it had been struck with the mattock, fell off; and, as the back of the head rested on the red-hot bottom bars of the furnace, the brains literally seethed, bubbled, and boiled as in a cauldron, for a very long time.

Byron could not face this scene, he withdrew to the beach and swam off to the *Bolivar*. Leigh Hunt remained in the carriage. The fire was so fierce as to produce a white heat on the iron, and to reduce its contents to grey ashes. The only portions that were not consumed were some fragments of bones, the jaw, and the skull, but what surprised us all, was that the heart remained entire. In snatching this relic from the fiery furnace, my hand was severely burnt.

EDWARD JOHN TRELAWNEY, *Recollections of the Last Days of Shelley and Byron*, 1858

On they went, singing 'Eternal Memory', and whenever they stopped, the sound of their feet, the horses and the gusts of wind seemed to carry on their singing.

Passers-by made way for the procession, counted the wreaths and crossed themselves. Some joined in out of curiosity and asked: 'Who is being buried?'—'Zhivago,' they were told.—'Oh, I see. That explains it.'—'It isn't him. It's his wife.'—'Well, it comes to the same thing. May she rest in peace. It's a fine funeral.'

The last moments flashed past, counted, irrevocable. 'The earth is the Lord's and the fullness thereof, the earth and everything that dwells therein.' The priest scattered earth in the form of a cross over the body of Marya Nikolayevna. They sang 'The souls of the just'. Then a fearful bustle began. The coffin was closed, nailed and lowered into the ground. Clods of earth drummed on the lid like rain as the grave was filled hurriedly by four spades. A mound grew up on it and a ten-year-old boy climbed on top.

Only the numb and unfeeling condition which comes to people at the end of a big funeral could account for some of the mourners' thinking that he wished to make an address over his mother's grave.

He raised his head and, from his vantage point, absently surveyed the bare autumn landscape and the domes of the monastery. His snub-nosed face was contorted. He stretched out his neck. If a wolf cub had done this it would have been obvious that it was about to howl. The boy covered his face with his hands and burst into sobs.

BORIS PASTERNAK (1890–1960), *Doctor Zhivago*, tr. Max Hayward and Manya Harari

Up yonder hill, behold how sadly slow
The bier moves winding from the vale below;
There lie the happy dead, from trouble free,
And the glad parish pays the frugal fee:
No more, O Death! thy victim starts to hear
Churchwarden stern, or kingly overseer;
No more the farmer claims his humble bow,
Thou art his lord, the best of tyrants thou!
  Now to the church behold the mourners come,
Sedately torpid and devoutly dumb;
The village children now their games suspend,
To see the bier that bears their ancient friend;
For he was one in all their idle sport,
And like a monarch ruled their little court.
The pliant bow he form'd, the flying ball,
The bat, the wicket, were his labours all;
Him now they follow to his grave, and stand,
Silent and sad, and gazing, hand in hand;
While bending low, their eager eyes explore
The mingled relics of the parish poor.
The bell tolls late, the moping owl flies round,
Fear marks the flight and magnifies the sound;
The busy priest, detain'd by weightier care,
Defers his duty till the day of prayer;
And, waiting long, the crowd retire distress'd,
To think a poor man's bones should lie unbless'd.

GEORGE CRABBE (1754–1832), *The Village* (a pauper's funeral)

Those graves, with bending osier bound,
That nameless heave the crumbled ground,
Quick to the glancing thought disclose
Where toil and poverty repose.

THOMAS PARNELL (1679–1718), from 'A Night-piece on Death'

The defunct was a young Union labourer, about twenty-five, who had been drowned the previous day, while trying to swim some horses across a billabong of the Darling . . .

The hearse was drawn up and the tail-boards were opened. The funeral extinguished its right ear with its hat as four men lifted the coffin out and laid it over the grave. The priest—a pale, quiet young fellow—stood under the shade of a sapling which grew at the head of the grave. He took off his hat, dropped it carelessly on the ground, and proceeded to business. I noticed that one or two heathens winced slightly when the holy water was sprinkled on the coffin. The drops quickly evaporated, and the little round black spots they left were soon dusted over; but the spots showed, by contrast, the cheapness and shabbiness of the cloth with which the coffin was covered. It seemed black before; now it looked a dusky grey . . .

The grave looked very narrow under the coffin, and I drew a breath of relief when the box slid easily down. I saw a coffin get stuck once, at Rookwood, and it had to be yanked out with difficulty, and laid on the sods at the feet of the heart-broken relations, who howled dismally while the grave-diggers widened the hole. But they don't cut contracts so fine in the West. Our grave-digger was not altogether bowelless, and, out of respect for that human quality described as 'feelin's', he scraped up some light and dusty soil and threw it down to deaden the fall of the clay lumps on the coffin. He also tried to steer the first few shovelfuls gently down against the end of the grave with the back of the shovel turned outwards, but the hard dry Darling River clods rebounded and knocked all the same. It didn't matter much—nothing does. The fall of lumps of clay on a stranger's coffin doesn't sound any different from the fall of the same things on an ordinary wooden box—at least I didn't notice anything awesome or unusual in the sound; but, perhaps, one of us—the most sensitive—might have been impressed by being reminded of a burial of long ago, when the thump of every sod jolted his heart.

I have left out the wattle—because it wasn't there. I have also neglected to mention the heart-broken old mate, with his grizzled head bowed and great pearly drops streaming down his rugged cheeks.

He was absent—he was probably 'out back'. For similar reasons I have omitted reference to the suspicious moisture in the eyes of a bearded bush ruffian named Bill. Bill failed to turn up, and the only moisture was that which was induced by the heat. I have left out the 'sad Australian sunset', because the sun was not going down at the time. The burial took place exactly at midday.

HENRY LAWSON (1867–1922), 'The Union Buries its Dead'

When we came to the chapel of Henry the Seventh, all solemnity and decorum ceased—no order was observed, people set or stood where they could or would, the yeomen of the guard were crying out for help, oppressed by the immense weight of the coffin, the Bishop read sadly, and blundered in the prayers, the fine chapter, *Man that is born of a woman*, was chanted, not read, and the anthem, besides being unmeasurably tedious, would have served as well for a nuptial. The real serious part was the figure of the Duke of Cumberland, heightened by a thousand melancholy circumstances. He had a dark brown adonis, and a cloak of black cloth, with a train of five yards. Attending the funeral of a father, however little reason he had so to love him, could not be pleasant. His leg extremely bad, yet forced to stand upon it near two hours, his face bloated and distorted with his late paralytic stroke, which has affected, too, one of his eyes, and placed over the mouth of the vault, into which, in all probability, he must himself so soon descend—think how unpleasant a situation! He bore it all with a firm and unaffected countenance. This grave scene was fully contrasted by the burlesque Duke of Newcastle. He fell into a fit of crying the moment he came into the chapel, and flung himself back in a stall, the Archbishop hovering over him with a smelling-bottle—but in two minutes his curiosity got the better of his hypocrisy, and he ran about the chapel with his glass to spy who was or was not there, spying with one hand, and mopping his eyes with t'other. Then returned the fear of catching cold, and the Duke of Cumberland, who was sinking with heat, felt himself weighed down, and turning round, found it was the Duke of Newcastle standing upon his train to avoid the chill of the marble. It was very theatric to look down into the vault, where the coffin lay, attended by mourners with lights. Clavering, the Groom of the Bedchamber, refused to sit up with the body, and was dismissed by the King's order.

HORACE WALPOLE, letter to George Montagu, 13 November 1760 (the funeral of George II)

Alas! England now mourns for her poet that's gone—
The late and the good Lord Tennyson.
I hope his soul has fled to heaven above,
Where there is everlasting joy and love.

He was a man that didn't care for company,
Because company interfered with his study,
And confused the bright ideas in his brain,
And for that reason from company he liked to abstain.

He has written some fine pieces of poetry in his time,
Especially the May Queen, which is really sublime;
Also the gallant charge of the Light Brigade—
A most heroic poem, and beautifully made.

He believed in the Bible, also in Shakespeare,
Which he advised young men to read without any fear;
And by following the advice of both works therein,
They would seldom or never commit any sin . . .

The chief mourners were all of the Tennyson family,
Including the Hon. Mr and Mrs Hallam Tennyson, and Masters Lionel
and Aubrey,
And Mr Arthur Tennyson, and Mr and Mrs Horatio Tennyson;
Also Sir Andrew Clark, who was looking woe begone.

The bottom of the grave was thickly strewn with white roses,
And for such a grave kings will sigh where the poet now reposes;
And many of the wreaths were much observed and commented upon,
And conspicuous amongst them was one from Mrs Gladstone.

The Gordon boys were there looking solemn and serene,
Also Sir Henry Ponsonby to represent the Queen;
Likewise Henry Irving, the great tragedian,
With a solemn aspect, and driving his brougham.

And, in conclusion, I most earnestly pray,
That the people will erect a monument for him without delay,
To commemorate the good work he has done,
And his name in gold letters written thereon!

WILLIAM MCGONAGALL (1830–1902), from 'Death and Burial of Lord Tennyson'

As regards notices of my death, you are to manage them between you. I would rather you sent out too few than too many. Those intended for people of high rank or position should be on light pink paper and the announcement should be in small letters. Uncoloured paper must not be used. For distribution to ordinary people, small, antique slips are the really refined thing. The sending out of large sheets of paper is a vulgarity . . .

As for recitation of Scriptures, chanting of liturgies and entertainment of monks on the seventh day—these are things that I have always detested. You may tell your sisters to come and make an offering to me, in which case I shall certainly accept it; or to come once and wail; at which I shall be deeply moved. But if monks come to the door, at the first sound of their wooden clappers, my divine soul will stop up its ears and run away, which I am sure you would not like.

The Will of Yuan Mei (1716–98), tr. Waley

A wedding-cake face in a paper frill.
How superior he is now.

It is like possessing a saint.
The nurses in their wing-caps are no longer so beautiful;

They are browning, like touched gardenias.
The bed is rolled from the wall.

This is what it is to be complete. It is horrible.
Is he wearing pajamas or an evening suit

Under the glued sheet from which his powdery beak
Rises so whitely unbuffeted?

They propped his jaw with a book until it stiffened
And folded his hands, that were shaking: goodbye, goodbye.

Now the washed sheets fly in the sun,
The pillow cases are sweetening.

It is a blessing, it is a blessing:
The long coffin of soap-coloured oak,

The curious bearers and the raw date
Engraving itself in silver with marvellous calm.

Sylvia Plath (1932–63), from 'Berck-Plage'

The coffin of the writer, so 'tenderly loved' by Moscow, was brought in a green wagon bearing the inscription 'Oysters' in big letters on the door. A section of the small crowd which had gathered at the station to meet the writer followed the coffin of General Keller just arrived from Manchuria, and wondered why Chekhov was being carried to his grave to the music of a military band. When the mistake was discovered certain genial persons began laughing and sniggering. Chekhov's coffin was followed by about a hundred people, not more. Two lawyers stand out in my memory, both in new boots and gaily patterned ties, like bridegrooms. Walking behind them I heard one of them, V. A. Maklakov, talking about the cleverness of dogs, and the other, whom I did not know, boasting of the convenience of his summer cottage and the beauty of its environments. And some lady in a purple dress, holding up a lace sunshade, was assuring an old gentleman in horn-rimmed spectacles: 'Oh, he was such a darling, and so witty . . .'

The old gentleman coughed incredulously. It was a hot, dusty day. The procession was headed by a stout police officer on a stout white horse.

MAXIM GORKY (1868–1936), *On Literature*, tr. Ivy Litvinov

The upper middles would probably drink themselves silly at the funeral. Although a few years ago this would have been frowned on. When my husband in the sixties announced that he intended to leave £200 in his will for a booze-up for his friends, his lawyer talked him out of it, saying it was in bad taste and would upset people. The same year his grandmother died, and after the funeral, recovering from the innate vulgarity of the cremation service when the gramophone record stuck on 'Abi-abi-abi-abi-de with me', the whole family trooped home and discovered some crates of Australian burgundy under the stairs. A rip-roaring party ensued and soon a lower middle busybody who lived next door came bustling over to see if anything was wrong. Whereupon my father-in-law, holding a glass and seeing her coming up the path, uttered the immortal line: 'Who is this intruding on our grief?'

JILLY COOPER, *Class*, 1979

'Thady, (says he) all you've been telling me brings a strange thought into my head; I've a notion I shall not be long for this world any how, and I've a great fancy to see my own funeral afore I die.' I was greatly shocked at the first speaking to hear him speak so light about his funeral, and he to all appearance in good health, but recollecting myself, answered—'To be sure it would be a fine sight as one could see, I dared to say, and one I should be proud to witness, and I did not doubt his honour's would be as great a funeral as ever Sir Patrick O'Shaughlin's was, and such a one as that had never been known in the county afore or since.' But I never thought he was in earnest about seeing his own funeral himself, till the next day he returns to it again.—'Thady, (says he) as far as the wake goes, sure I might without any great trouble have the satisfaction of seeing a bit of my own funeral.'—'Well, since your honour's honour's so bent upon it, (says I, not willing to cross him, and he in trouble) we must see what we can do.'—So he fell into a sort of a sham disorder, which was easy done, as he kept his bed and no one to see him; and I got my shister, who was an old woman very handy about the sick, and very skilful, to come up to the Lodge to nurse him; and we gave out, she knowing no better, that he was just at his latter end, and it answered beyond any thing; and there was a great throng of people, men, women and childer, and there being only two rooms at the Lodge, . . . the house was soon as full and fuller than it could hold, and the heat, and smoke, and noise wonderful great; and standing amongst them that were near the bed, but not thinking at all of the dead, I was started by the sound of my master's voice from under the great coats that had been thrown all at top, and I went close up, no one noticing.—'Thady, (says he) I've had enough of this, I'm smothering, and I can't hear a word of all they're saying of the deceased.'—'God bless you, and lie still quiet (says I) a bit longer, for my shister's afraid of ghosts, and would die on the spot with the fright, was she to see you come to life all on a sudden this way without the least preparation.'—So he lays him still, though well nigh stifled, and I made all haste to tell the secret of the joke, whispering to one and t'other, and there was a great surprise, but not so great as we had laid out it would.—'And aren't we to have the pipes and tobacco, after coming so far tonight?' says some; but they were all well enough pleased when his honour got up to drink with them, and sent for more spirits from a shebean-house, where they very civilly let him have it upon credit—so the night passed off very merrily, but to my mind Sir Condy was rather upon the sad order in the midst of it all, not finding there had been such a great talk about himself after his death as he had always expected to hear.

MARIA EDGEWORTH (1767–1849), *Castle Rackrent*

At length the day of the funeral, pious and truthful ceremony that it was, arrived. Mr Mould, with a glass of generous port between his eye and the light, leaned against the desk in the little glass office with his gold watch in his unoccupied hand, and conversed with Mrs Gamp; two mutes were at the house-door, looking as mournful as could be reasonably expected of men with such a thriving job in hand; the whole of Mr Mould's establishment were on duty within the house or without; feathers waved, horses snorted, silk and velvets fluttered; in a word, as Mr Mould emphatically said, 'everything that money could do was done.'

'And what can do more, Mrs Gamp?' exclaimed the undertaker, as he emptied his glass and smacked his lips.

'Nothing in the world, sir.'

'Nothing in the world,' repeated Mr Mould. 'You are right, Mrs Gamp. Why do people spend more money': here he filled his glass again: 'upon a death, Mrs Gamp, than upon a birth? Come, that's in your way; you ought to know. How do you account for that now?'

'Perhaps it is because an undertaker's charges comes dearer than a nurse's charges, sir,' said Mrs Gamp, tittering, and smoothing down her new black dress with her hands . . .

'No, Mrs Gamp; I'll tell you why it is. It's because the laying out of money with a well-conducted establishment, where the thing is performed upon the very best scale, binds the broken heart, and sheds balm upon the wounded spirit. Hearts want binding, and spirits want balming when people die: not when people are born. Look at this gentleman today; look at him.'

'An open-handed gentleman?' cried Mrs Gamp, with enthusiasm.

'No, no,' said the undertaker; 'not an open-handed gentleman in general, by any means. There you mistake him: but an afflicted gentleman, an affectionate gentleman, who knows what it is in the power of money to do, in giving him relief, and in testifying his love and veneration for the departed. It can give him,' said Mr Mould, waving his watch-chain slowly round and round, so that he described one circle after every item; 'it can give him four horses to each vehicle; it can give him velvet trappings; it can give him drivers in cloth cloaks and top-boots; it can give him the plumage of the ostrich, dyed black; it can give him any number of walking attendants, dressed in the first style of funeral fashion, and carrying batons tipped with brass; it can give him a handsome tomb; it can give him a place in Westminster Abbey itself, if he choose to invest it in such a purchase. Oh! do not let us say that gold is dross, when it can buy such things as these, Mrs Gamp.'

CHARLES DICKENS (1812–70), *Martin Chuzzlewit*

The hearse comes up the road
With its funeral load

Sharp on the stroke of twelve.
I greet it myself,

Good-morning the head man
Who's brought the dead man.

I say we're four only.
Still, he won't be lonely.

Being next of kin
I'm the first one in

Behind the bearers,
The black mourning wearers.

(A quick thought appals:
What if one trips and falls?)

They lay him safely down,
The coffin a light brown.

Prayers begin. I sit
And let my mind admit

That screwed-down speechless thing
And how another spring

His spouse was carried here.
Now they're remarried here

And may be happier even
In the clean church of heaven.

We say the last amen.
A button's pressed and then

To canned funeral strains
His dear dead remains,

Eighty-four years gone by,
Sink with a whirring sigh.

I tip and say goodbye.

J. C. HALL (b. 1920), 'Twelve Minutes'

So that this too, too solid flesh might not melt, we are offered 'solid copper—a quality casket which offers superb value to the client seeking long-lasting protection', or 'the Colonial Classic Beauty—18 gauge lead coated steel, seamless top, lap-jointed welded body construction'. Some are equipped with foam rubber, some with inner-spring mattresses. Elgin offers 'the revolutionary "Perfect-Posture" bed'. Not every casket need have a silver lining, for one may choose between 'more than 60 colour matched shades, magnificent and unique masterpieces' by the Cheney casket-lining people. Shrouds no longer exist. Instead, you may patronize a grave-wear couturière who promises 'handmade original fashions—styles from the best in life for the last memory—dresses, men's suits, negligees, accessories'. For the final, perfect grooming: 'Nature-Glo—the ultimate in cosmetic embalming'. And, where have we heard that phrase 'peace of mind protection' before? No matter. In funeral advertising, it is applied to the Wilbert Burial Vault, with its $\frac{3}{8}$-inch precast asphalt inner liner plus extra-thick, reinforced concrete—all this 'guaranteed by Good Housekeeping'. Here again the Cadillac, status symbol par excellence, appears in all its gleaming glory, this time transformed into a pastel-coloured funeral hearse.

JESSICA MITFORD, *The American Way of Death*, 1963

Sir, I recently helped a friend make funeral arrangements for a deceased relative and was horrified to learn that the cost of the cheapest coffin (not the funeral) was £286. As my friend's relative was to be cremated, presumably the coffin also would be cremated?

As I approach my three score years and ten, would it not be a good investment to buy my coffin now.

Yours faithfully,
ELISABETH GOODWIN

Sir, An excellent suggestion has been made by Mrs Goodwin in her letter today (January 28) that older people should buy their coffins now, a practice incidentally followed by many Chinese for centuries.

The chief difficulty, however, would be one of storage in these days of many flat-dwellers. Would my visitors be elated or depressed, I wonder, by the sight of a coffin propped up in a small entrance hall?

Yours faithfully,
MARGARET I. KILLERY

Sir, If nothing else the British are innovators. Mrs Killery (January 30) need have no fear that the coffin propped up in the entrance hall would cause concern, for the convertible coffin poses only a passing challenge to our national ingenuity. From cocktail cabinet to cloak cupboard the range is infinite.

Here the horologist comes into his own. With the sweeping sickle of the second hand and the automated hour-glass finial, who can doubt that a convertible grandfather clock would prove to be the most popular and appropriate retirement present of all?

Yours faithfully,

PATRICK H. KEMP

Letters to *The Times*, 1981

I telephoned one of the large San Francisco dailies, and was connected with Miss Black . . .

'The notice is to read "Please omit flowers",' I said.

'Well, we never put it that way. How about "Memorial Contributions Preferred", or "Memorial gifts . . ."?' suggested Miss Black.

'No, I don't think that will do. The family wouldn't like to ask for charitable contributions on an occasion like this. Besides, my friend's mother left exact instructions in her will about the wording of the notice, and she specified that it should say, "Please omit flowers".'

Sounds of Miss Black being in deep water. 'I'm sorry, ma'am, that would be against our policy. We are not allowed to accept ads that are derogatory about anyone, or about any*thing*.'

'But this isn't derogatory about anyone.'

'It's derogatory about flowers.'

'There is nothing unkind about flowers in that notice. As a matter of fact, my friend's mother adored flowers. She just doesn't want them cluttering up her funeral, that's all.'

Miss Black was firm; she spoke about newspaper policy, she said she had her instructions. I was firm; I spoke about freedom of the press and the rights of the individual, but to no avail. Later, I telephoned the head of the department; he was firm, too. 'We couldn't publish a notice like that,' he said. 'Why, the florists would be right on our necks!'

JESSICA MITFORD, op. cit.

Dr Eaton has set up his Credo at the entrance. 'I believe in a happy Eternal Life,' he says. 'I believe those of us left behind should be glad in the certain belief that those gone before have entered into that happier Life.' This theme is repeated on Coleus Terrace: 'Be happy because they for whom you mourn are happy—far happier than ever before.' And again in Vesperland: '. . . Happy because Forest Lawn has eradicated the old customs of Death and depicts Life not Death.'

The implication of these texts is clear. Forest Lawn has consciously turned its back on the 'old customs of death', the grim traditional alternatives of Heaven and Hell, and promises immediate eternal happiness for all its inmates. Similar claims are made for other holy places—the Ganges, Debra Lebanos in Abyssinia, and so on. Some of the simpler crusaders probably believed that they would go straight to Heaven if they died in the Holy Land. But there is a catch in most of these dispensations, a sincere repentance, sometimes an arduous pilgrimage, sometimes a monastic rule in the closing years. Dr Eaton is the first man to offer eternal salvation at an inclusive charge as part of his undertaking service.

EVELYN WAUGH (1903–66), 'Half in Love with Easeful Death'

In 1887, when Villiers de l'Isle-Adam and Léon Bloy were passing the flower-sellers, monumental masons and shops specializing in funeral accessories near the Père Lachaise cemetery, Villiers exclaimed in fury: 'Those are the people who invented death!'

v. A. W. RAITT, *The Life of Villiers de l'Isle-Adam*, 1981

Now, had Tashtego perished in that head [a sperm whale's], it had been a very precious perishing; smothered in the very whitest and daintiest of fragrant spermaceti; coffined, hearsed, and tombed in the secret inner chamber and sanctum sanctorum of the whale. Only one sweeter end can readily be recalled—the delicious death of an Ohio honey-hunter, who seeking honey in the crotch of a hollow tree, found such exceeding store of it, that leaning too far over, it sucked him in, so that he died embalmed. How many, think ye, have likewise fallen into Plato's honey head, and sweetly perished there?

HERMAN MELVILLE (1819–91), *Moby Dick*

Los Angeles, June 7—Robert Nelson, the former president of a society which froze bodies to await a day when science found a way of restoring life, and Joseph Klockgether, an undertaker, were ordered today to pay nearly $1m (£513,000) in damages.

Relatives of the dead had filed a lawsuit alleging fraud. They claimed that the corpses, which were put in capsules to await a scientific breakthrough, were not kept in a perpetually frozen state.

The Times, 1981

To be knaved out of our graves, to have our skulls made drinking-bowls, and our bones turned into Pipes, to delight and sport our Enemies, are Tragical abominations escaped in burning Burials.

BROWNE, op. cit.

'Tell me, good dog, whose tomb you guard so well.'
'The Cynic's.' 'True; but who that Cynic tell.'
'Diogenes, of fair Sinope's race.'
'What? He that in a tub was wont to dwell?'
'Yes: but the stars are now his dwelling-place.'

ANON., tr. John Addington Symonds

Resurrections and Immortalities

THE passages that follow outline the great debate on immortality, from the Socratic arguments and the glorious visions of Donne and other believers in an afterlife, Christian or otherwise, to the worldly wisdom of Swift as he contemplates with horror the Struldbruggs' deprivation of death, and to those who more mildly consider that one existence is enough and the rest can well be silence and peace.

Nabokov was not the first to formulate the paradox whereby we are content to deny immortality to other people while keeping it for ourselves. More than two hundred years earlier Edward Young wrote in *Night Thoughts on Life, Death and Immortality*, a poem less lugubrious than its reputation would have: 'All men think all men mortal, but themselves.' In an extract included here Freud proposes a shrewd answer to the riddle: in visualizing our own death we play the part of spectators—that is, we are undeniably alive—and hence, in the unconscious, we are implicitly persuaded of our own immortality. Perhaps this is to credit the unconscious with too much influence and the conscious mind with too little versatility.

Fair space has been given to secular varieties of immortality: the superb solipsism of Goethe, Keats's nightingale ('no hungry generations tread thee down'), Proust's moving vision of Bergotte's books keeping vigil in the shop-windows, and the more humble or ironic ambitions of Ringelnatz and Mandelstam. This conception of survival is epitomized with considerable solemnity in George Eliot's poem; in her case and in some others the hope of living on usefully in the minds of later generations has been duly fulfilled.

Since they normally tend to be reticent about other-worldly matters, it is not surprising that little is heard from modern theologians and ecclesiastics. But the persistence of man's desire or need to look forward to immortality in some form, against much of the evidence and yet (it is held) in accordance with some of it, is demonstrated in a further passage from the pragmatic William James's book, *The Will to Believe*, where he speaks of the existence of an invisible world which may in part depend upon us, our responses and efforts. Our present life, he writes, '*feels* like a real fight—as if there were something really wild in the universe which we, with all our idealities and faithfulnesses, are needed to redeem; and first of all to redeem our own hearts from atheisms and fears. For such a half-wild, half-saved universe our nature is adapted . . .' He concludes, rather perfunctorily: 'Believe that life *is* worth living, and your belief will help create the fact.' The same desire or need informs Unamuno's equally tentative but more intense speculations: even in the non-believer there lurks a vague shadow, 'a shadow of the shadow of uncertainty', in some inner recess of his spirit. 'How, without this uncertainty, could we ever live?' That famous will to survive does not give up easily.

~

And, behold, two of them went that same day to a village called Emmaus, which was from Jerusalem about threescore furlongs. And they talked together of all these things which had happened. And it came to pass, that, while they communed together and reasoned, Jesus himself drew near, and went with them. But their eyes were holden that they should not know him. And he said unto them, What manner of communications are these that ye have one to another, as ye walk, and are sad? And the one of them, whose name was Cleopas, answering said unto him, Art thou only a stranger in Jerusalem, and hast not known the things which are come to pass there in these days?

And he said unto them, What things? And they said unto him, Concerning Jesus of Nazareth, which was a prophet mighty in deed and word before God and all the people: and how the chief priests and our rulers delivered him to be condemned to death, and have crucified him. But we trusted that it had been he which should have redeemed Israel: and beside all this, today is the third day since these things were done . . .

Then he said unto them, O fools, and slow of heart to believe all that the prophets have spoken: Ought not Christ to have suffered these things, and to enter into his glory? And beginning at Moses and all the prophets, he expounded unto them in all the scriptures the things concerning himself.

And they drew nigh unto the village, whither they went: and he made as though he would have gone further. But they constrained him, saying, Abide with us: for it is toward evening, and the day is far spent. And he went in to tarry with them.

And it came to pass, as he sat at meat with them, he took bread, and blessed it, and brake, and gave to them. And their eyes were opened, and they knew him; and he vanished out of their sight.

St Luke, 24

Dear, beauteous death! the Jewel of the Just,
  Shining no where, but in the dark;
What mysteries do lie beyond thy dust,
  Could man outlook that mark!

He that hath found some fledg'd bird's nest, may know
  At first sight, if the bird be flown;
But what fair Well, or Grove he sings in now,
  That is to him unknown.

And yet, as Angels in some brighter dreams
  Call to the soul, when man doth sleep:
So some strange thoughts transcend our wonted themes,
  And into glory peep . . .

HENRY VAUGHAN (*c.*1622–95), from 'Ascension Hymn'

'Tis but a night, a long and moonless night,
We make the *Grave* our bed, and then are gone.
Thus at the shut of ev'n, the weary bird
Leaves the wide air, and in some lonely brake
Cow'rs down, and dozes till the dawn of day,
Then claps his well-fledg'd wings, and bears away.

ROBERT BLAIR (1699–1746), *The Grave*

Man that is born of a woman is of few days, and full of trouble. He cometh forth like a flower, and is cut down: he fleeth also as a shadow, and continueth not.

For there is hope of a tree, if it be cut down, that it will sprout again, and that the tender branch thereof will not cease. Though the root thereof wax old in the earth, and the stock thereof die in the ground; Yet through the scent of water it will bud, and bring forth boughs like a plant. But man dieth, and wasteth away: yea, man giveth up the ghost, and where is he? As the waters fail from the sea, and the flood decayeth and drieth up: So man lieth down, and riseth not: till the heavens be no more, they shall not awake, nor be raised out of their sleep. O that thou wouldest hide me in the grave, that thou wouldest keep me secret, until thy wrath be past, that thou wouldest appoint me a set time, and remember me! If a man die, shall he live again? all the days of my appointed time will I wait, till my change come.

Job, 14

Verily, verily, I say unto you, Except a corn of wheat fall into the ground and die, it abideth alone; but if it die, it bringeth forth much fruit.

St John, 12

At the round earth's imagin'd corners, blow
Your trumpets, Angels, and arise, arise
From death, you numberless infinities
Of souls, and to your scatter'd bodies go,
All whom the flood did, and fire shall o'erthrow,
All whom war, dearth, age, agues, tyrannies,
Despair, law, chance, hath slain, and you whose eyes
Shall behold God, and never taste death's woe.
But let them sleep, Lord, and me mourn a space,
For, if above all these, my sins abound,
'Tis late to ask abundance of thy grace,
When we are there; here on this lowly ground,
Teach me how to repent; for that's as good
As if thou'hadst seal'd my pardon, with thy blood.

DONNE, *Holy Sonnets*

Come away,
Make no delay.
Summon all the dust to rise,
Till it stir, and rub the eyes;
While this member jogs the other,
Each one whisp'ring, '*Live you, brother?*' . . .

Come away,
Help our decay.
Man is out of order hurl'd,
Parcel'd out to all the world.
Lord, thy broken consort raise,
And the music shall be praise.

GEORGE HERBERT (1593–1633), from 'Dooms-day'

On the day of death, when my bier is on the move, do not suppose that I have any pain at leaving this world.

Do not weep for me, say not 'Alas, alas!' You will fall into the devil's snare—that would indeed be alas!

When you see my hearse, say not 'Parting, parting!' That time there will be for me union and encounter.

When you commit me to the grave, say not 'Farewell, farewell!' For the grave is a veil over the reunion of paradise.

Having seen the going-down, look upon the coming-up; how should setting impair the sun and the moon?

To you it appears as setting, but it is a rising; the tomb appears as a prison, but it is release for the soul.

What seed ever went down into the earth which did not grow? Why do you doubt so regarding the human seed?

What bucket ever went down and came not out full? Why this complaining of the well by the Joseph of the spirit?

When you have closed your mouth on this side, open it on that, for your shout of triumph will echo in the placeless air.

JALĀL AL-DĪN RŪMĪ (1207–73), tr. A. J. Arberry

'And with what body do they come?'—
Then they *do* come—Rejoice!
What Door—What Hour—Run—run—My Soul!
Illuminate the House!

'Body!' Then real—a Face and Eyes—
To know that it is them!—
Paul knew the Man that knew the News—
He passed through Bethlehem—

EMILY DICKINSON, 'And with what body do they come?'

I well remember how one no longer present with us, but to whom I cease not to look up, shrank from entering the Mummy Room at the British Museum under a vivid realization of how the general resurrection might occur even as one stood among those solemn corpses turned into a sight for sightseers.

And at that great and awful day, what will be thought of supposititious heads and members?

CHRISTINA ROSSETTI, *Time Flies*

From Minneapolis and Rio, from Sydney and Hendon South,
they will tunnel through the soil
(old mole in *Hamlet* and the kids' cartoons)
and eventually they'll end up here.
This is the Jewish Cemetery at the foot of
The Mount of Olives. From my point of view
(and I have been calling myself a residual Christian),
a cemetery without crosses and headstones
looks like a stonemason's junkyard. But
just beyond the cemetery is the Garden
of Gethsemane, the Vale of Kidron, the Dome
of the Rock, and the Eastern Wall
with its Gate for the Messiah to come through
(decently bricked-up). Staring in sunlight,
I am conscious only of the Jewish Resurrection—
Christians can resurrect anywhere,
but every Jew has to go by underground:
out of Cracow or Lodz, with his diary's
death, from the musical comedy stage of
New York, from the pages of fiery
Medieval books—he dives into oblivion
and comes up here to join the queues
as the millions swapping stories enter Heaven.
How hot and dirty and pale are olive leaves,
how one's heart beats above the golden city,
which really belongs to architects,
archaeologists, and such Sunday Schools as we have left.
None of it's as real as a carved camel
or a smudgy glass ring, except this Resurrection:
I can see this easily, and if the boys
flogging panoramic views would just keep quiet
I could hear the first tapping on a plate.
Now we are changed: I haven't an atom
in my body which I brought to Europe
in 1951. How beautiful is evolution,
that I am moving deeper into my own brain.
The air is filled with something
and I will not call it light—
stay with me, my friends; truth and love,
like miracles, need nowhere at all to happen in.

PETER PORTER (b. 1929), 'Evolution'

At one point we got on the theme of immortality, in which she believed without being sure of its precise form. 'There is no death,' she said. 'No, my dear lady, but there are funerals.'

PETER DE VRIES (b. 1910), *Comfort Me With Apples*

Mr Kernan said with solemnity:

—*I am the resurrection and the life*. That touches a man's inmost heart.

—It does, Mr Bloom said.

Your heart perhaps but what price the fellow in the six feet by two with his toes to the daisies? No touching that. Seat of the affections. Broken heart. A pump after all, pumping thousands of gallons of blood every day. One fine day it gets bunged up and there you are. Lots of them lying around here: lungs, hearts, livers. Old rusty pumps: damn the thing else. The resurrection and the life. Once you are dead you are dead. That last day idea. Knocking them all up out of their graves. Come forth, Lazarus! And he came fifth and lost the job. Get up! Last day! Then every fellow mousing around for his liver and his lights and the rest of his traps. Find damn all of himself that morning. Pennyweight of powder in a skull. Twelve grammes one pennyweight. Troy measure.

JOYCE, *Ulysses*

Out of the tomb, we bring Badroulbadour,
Within our bellies, we her chariot.
Here is an eye. And here are, one by one,
The lashes of that eye and its white lid.
Here is the cheek on which that lid declined,
And, finger after finger, here, the hand,
The genius of that cheek. Here are the lips,
The bundle of the body and the feet.

* * *

Out of the tomb we bring Badroulbadour.

WALLACE STEVENS (1879–1955), 'The Worms at Heaven's Gate'

It is indeed impossible to imagine our own death; and whenever we attempt to do so we can perceive that we are in fact still present as spectators. Hence the psycho-analytic school could venture on the assertion that at bottom no one believes in his own death, or, to put the same thing in another way, that in the unconscious every one of us is convinced of his own immortality.

FRÉUD, 'Thoughts for the Times on War and Death', tr. James Strachey, *The Standard Edition of the Complete Psychological Works of Sigmund Freud* (Volume XIV)

What moment in the gradual decay
Does resurrection choose? What year? What day?
Who has the stopwatch? Who rewinds the tape?
Are some less lucky, or do all escape?
A syllogism: *other men die; but I*
Am not another; therefore I'll not die.

VLADIMIR NABOKOV (1899–1977), *Pale Fire*

Those who wish to abolish death (whether by physical or metaphysical means)—at what stage of life do they want the process to be halted? At the age of twenty? At thirty-five, in our prime? To be thirty-five for two years sounds attractive, certainly. But for three years? A little dull, surely. For five years—ridiculous. For ten—tragic.

The film is so absorbing that we want this bit to go on and on . . .

You mean, you want the projector stopped, to watch a single motionless frame? No, no, no, but . . . Perhaps you'd like the whole sequence made up as an endless band, and projected indefinitely? Not that, either.

The sea and the stars and the wastes of the desert go on forever, and will not die. But the sea and the stars and the wastes of the desert are dead already.

MICHAEL FRAYN (b. 1933), *Constructions*

This night as I was in my sleep, I dreamed, and behold the heavens grew exceedingly black; also it thundered and lightened in such fearful wise, that it put me into an agony. So I looked up in my dream, and saw the clouds racked at an unusual rate; upon which, I heard a great sound of a trumpet, and saw also a man sit upon a cloud, attended with the thousands of Heaven: they were all in flaming fire; also the heavens were in a burning flame. I heard then a voice, saying, 'Arise, ye dead, and come to judgement'; and with that, the rocks rent, the graves opened, and the dead that were therein, came forth;

some of them were exceeding glad, and looked upward: some sought to hide themselves under the mountains; then I saw the man, that sat upon the cloud, open the book, and bid the world draw near. Yet there was, by reason of a fierce flame which issued out and came from before him, as convenient a distance betwixt him and them, as betwixt the judge and the prisoners at the bar. I heard it also proclaimed to them that attended on the man that sat on the cloud, 'Gather together the tares, the chaff, and stubble, and cast them into the burning lake': and with that, the bottomless pit opened, just whereabouts I stood; out of the mouth of which there came, in an abundant manner, smoke and coals of fire, with hideous noises. It was also said to the same persons, 'Gather up my wheat into the garner.' And with that, I saw many catched up and carried away into the clouds.

JOHN BUNYAN (1628–88), *The Pilgrim's Progress*

That day of wrath, that dreadful day,
When heaven and earth shall pass away,
What power shall be the sinner's stay?
How shall he meet that dreadful day?

When, shrivelling like a parchèd scroll,
The flaming heavens together roll;
When louder yet, and yet more dread,
Swells the high trump that wakes the dead;

Oh! on that day, that wrathful day,
When man to judgement wakes from clay,
Be THOU the trembling sinner's stay,
Though heaven and earth shall pass away!

'Dies Irae' (13th century): Sir Walter Scott's version, *The Lay of the Last Minstrel*, 1805

At the round earth's imagined corners let
Angels regale us with a brass quartet,
Capping that concord with a fourfold shout:
'Out, everybody, everybody out!'
Then skeletons will rattle all about
Forming in file, on all fours, tail to snout,
Putting on flesh and face until they get,
Upright, to where the Judgement Seat is set.

There the All High, maternal, systematic,
Will separate the black souls from the white:
That lot there for the cellar, this the attic.
The wing'd musicians now will chime or blare a
Brief final tune, then they'll put out the light:
Er-phwhoo.
And so to bed.
Owwwwwww.
Bona sera.

GIUSEPPE BELLI (1791–1863), 'The Last Judgement', tr. Anthony Burgess

It is true of the body, is it not? that physical evil, namely disease, wastes and destroys it until it is no longer a body at all, and all the other things we instanced are annihilated by the pervading corruption of the evil which peculiarly besets them. Now is it true in the same way of the soul that injustice and other forms of vice, by besetting and pervading it, waste it away in corruption until they sever it from the body and bring about its death?

No, certainly not . . .

For if its own evil and depravity cannot kill the soul, it is hardly likely that an evil designed for the destruction of a different thing will destroy the soul or anything but its own proper object. So, since the soul is not destroyed by any evil, either its own or another's, clearly it must be a thing that exists for ever, and is consequently immortal.

PLATO, *The Republic*, tr. F. M. Cornford

Behold, I show you a mystery; We shall not all sleep, but we shall all be changed, In a moment, in the twinkling of an eye, at the last trump: for the trumpet shall sound, and the dead shall be raised incorruptible, and we shall be changed. For this corruptible must put on incorruption, and this mortal must put on immortality. So when this corruptible shall have put on incorruption, and this mortal shall have put on immortality, then shall be brought to pass the saying that is written, Death is swallowed up in victory. O death, where is thy sting? O grave, where is thy victory?

I Corinthians, 15

What reason do atheists have to say that one cannot rise from the dead? Which is the more difficult, to be born or to be reborn? That

that which has never existed should exist, or that that which has existed should exist again? Is it more difficult to come into being than to return to it? Custom makes the one seem easy, absence of custom makes the other seem impossible: a vulgar way of judging!

PASCAL, *Pensées*

Socrates proved the immortality of the soul from the fact that sickness of the soul (sin) does not consume it as sickness of the body consumes the body. Similarly, the eternal in a person can be proved by the fact that despair cannot consume his self, that precisely this is the torment of contradiction in despair. If there were nothing eternal in a man, he could not despair at all; if despair could consume his self, then there would be no despair at all.

SØREN KIERKEGAARD (1813–55), *The Sickness unto Death*, tr. Howard V. Hong and Edna H. Hong

Sweet day, so cool, so calm, so bright,
The bridal of the earth and sky:
The dew shall weep thy fall tonight;
For thou must die.

Sweet rose, whose hue angry and brave
Bids the rash gazer wipe his eye:
Thy root is ever in its grave,
And thou must die.

Sweet spring, full of sweet days and roses,
A box where sweets compacted lie;
My music shows ye have your closes,
And all must die.

Only a sweet and virtuous soul,
Like season'd timber, never gives;
But though the whole world turns to coal,
Then chiefly lives.

HERBERT, 'Virtue'

A scientist risks his life for a new discovery in the realm of matter, a pioneer to establish a new settlement, an aviator to improve our means of communication, a miner to extract coal from the earth, a pearl fisher to filch from the ocean an ornament for the beauty of some unknown woman, a traveller to contemplate new landscapes, a mountain climber to conquer a bit of earth. What comparison is there between the result to be obtained, be it momentous or slight, and the price of human life which is thus wagered, the value of that being, full of promise, endowed with so many gifts and whom many hearts may love? Well, at each corner of human activity death lies in ambush. Every day we trust our lives and those of our beloved to the unknown driver of a subway train, of a plane, of a bus or a taxi. Where there is no risk, there is no life. A wisdom or a civilization based on the avoidance of risk, by virtue of a misinterpretation of the value of the human being, would run the greatest of all risks, that of cowardice and of deadly stupidity. That perpetual risk which man takes is the very condition of his life. That squandering of the human being is a law of nature; it is also the proof of the confidence, the trust and the elementary love we give every day to the divine principle from which we proceed, the very law of which is superabundance and generosity.

Now we face a paradox: on the one hand nothing in the world is more precious than one single human person; on the other hand nothing in the world is more squandered, more exposed to all kinds of dangers, than the human being—and this condition must be. What is the meaning of this paradox? It is perfectly clear. We have here a sign that man knows very well that death is not an end, but a beginning. He knows very well, in the secret depths of his own being, that he can run all risks, spend his life and scatter his possessions here below, because he is immortal. The chant of the Christian liturgy before the body of the deceased is significant: Life is changed, life is not taken away.

JACQUES MARITAIN (1882–1973), *Man's Destiny in Eternity*

Total annihilation is impossible. We are the prisoners of an infinity without outlet, wherein nothing perishes, wherein everything is dispersed, but nothing lost. Neither a body nor a thought can drop out of the universe, out of time and space. Not an atom of our flesh, not a quiver of our nerves, will go where they will cease to be, for there is no place where anything ceases to be. The brightness of a star extinguished millions of years ago still wanders in the ether where our eyes will perhaps behold it this very night, pursuing its endless road. It is the same with all that we see, as with all that we do not see. To be able to do away with a thing, that is to say, to fling it into nothingness, nothingness would have to exist; and, if it exist, under whatever form, it is no longer nothingness. As soon as we try to analyse it, to define it, or to understand it, thoughts and expressions fail us, or create that which they are struggling to deny. It is as contrary to the nature of our reason and probably of all imaginable reason to conceive nothingness as to conceive limits to infinity. Nothingness, besides, is but a negative infinity, a sort of infinity of darkness opposed to that which our intelligence strives to enlighten, or rather it is but a child-name or nickname which our mind has bestowed upon that which it has not attempted to embrace, for we call nothingness all that which escapes our senses or our reason and exists without our knowledge. The more that human thought rises and increases, the less comprehensible does nothingness become. In any case—and this is what matters here—if nothingness were possible, since it could not be anything whatever, it could not be dreadful.

MAURICE MAETERLINCK (1862–1949), *Death*, tr. Alexander Teixeira de Mattos

The absolute certainty that death is a complete and definitive and irrevocable annihilation of personal consciousness, a certainty of the same order as our certainty that the three angles of a triangle are equal to two right angles, or, contrariwise, the absolute certainty that our personal consciousness continues beyond death in whatever condition (including in such a concept the strange and adventitious additional notion of eternal reward or punishment)—either of these certainties would make our life equally impossible. In the most secret recess of the spirit of the man who believes that death will put an end to his personal consciousness and even to his memory forever, in that inner recess, even without his knowing it perhaps, a shadow hovers, a vague shadow lurks, a shadow of the shadow of uncertainty, and, while he tells himself: 'There's nothing for it but to live this passing life, for there is no other!' at the same time he hears, in this most secret recess, his own doubt murmur: 'Who knows? . . .' He is not sure he hears aright, but he hears. Likewise, in some recess of the soul of the true believer who has faith in a future life, a muffled voice, the voice of uncertainty, murmurs in his spirit's ear: 'Who knows? . . .' Perhaps these voices are no louder than the buzzing of mosquitoes when the wind roars through the trees in the woods; we scarcely make out the humming, and yet, mingled in the uproar of the storm, it can be heard. How, without this uncertainty, could we ever live?

MIGUEL DE UNAMUNO (1864–1936), *The Tragic Sense of Life*, tr. Anthony Kerrigan

Suppose, then, that after the greatest, most passionate vividness and tender glory, oblivion is all we have to expect, the big blank of death. What options present themselves? One option is to train yourself gradually into oblivion so that no great change has taken place when you have died. Another option is to increase the bitterness of life so that death is a desirable release. (In this the rest of mankind will fully collaborate.) There is a further option seldom chosen. That option is to let the deepest elements in you disclose their deepest information. If there is nothing but nonbeing and oblivion waiting for us, the prevailing beliefs have not misled us, and that's that. This would astonish me, for the prevailing beliefs seldom satisfy my need for truth. Still the possibility must be allowed.

SAUL BELLOW (b. 1915), *Humboldt's Gift*

One day in much good company, I was asked by a person of quality, whether I had seen any of their Struldbruggs, or Immortals . . . He said they commonly acted like mortals, until about thirty years old, after which by degrees they grew melancholy and dejected, increasing in both until they came to fourscore . . . When they came to fourscore years, which is reckoned the extremity of living in this country, they had not only all the follies and infirmities of other old men, but many more which arose from the dreadful prospect of never dying. They were not only opinionative, peevish, covetous, morose, vain, talkative; but uncapable of friendship, and dead to all natural affection, which never descended below their grandchildren. Envy and impotent desires, are their prevailing passions. But those objects against which their envy seems principally directed, are the vices of the younger sort, and the deaths of the old. By reflecting on the former, they find themselves cut off from all possibility of pleasure; and whenever they see a funeral, they lament and repine that others are gone to a harbour of rest, to which they themselves never can hope to arrive. They have no remembrance of any thing but what they learned and observed in their youth and middle age, and even that is very imperfect . . . The least miserable among them, appear to be those who turn to dotage, and entirely lose their memories; these meet with more pity and assistance, because they want many bad qualities which abound in others.

If a Struldbrugg happen to marry one of his own kind, the marriage is dissolved of course by the courtesy of the kingdom, as soon as the younger of the two comes to be fourscore. For the law thinks it a reasonable indulgence, that those who are condemned without any fault of their own, to a perpetual continuance in the world, should not have their misery doubled by the load of a wife.

As soon as they have completed the term of eighty years, they are looked on as dead in law; their heirs immediately succeed to their estates, only a small pittance is reserved for their support; and the poor ones are maintained at the public charge . . .

At ninety they lose their teeth and hair; they have at that age no distinction of taste, but eat and drink what ever they can get, without relish or appetite. The diseases they were subject to, still continue without increasing or diminishing. In talking, they forget the common appellation of things, and the names of persons, even of those, who are their nearest friends and relations. For the same reason, they never can amuse themselves with reading, because their memory will not serve to carry them from the beginning of a sentence to the end; and by this defect, they are deprived of the only entertainment whereof they might otherwise be capable . . .

The reader will easily believe, that from what I had heard and seen, my keen appetite for perpetuity of life was much abated. I grew heartily ashamed of the pleasing visions I had formed; and thought no tyrant could invent a death, into which I would not run with pleasure from such a life.

SWIFT, *Gulliver's Travels*

Most people think they have stuff in them for greater things than time suffers them to perform. To imagine a second career is a pleasing antidote for ill-fortune; the poor soul wants another chance. But how should a future life be constituted if it is to satisfy this demand, and how long need it last? It would evidently have to go on in an environment closely analogous to earth; I could not, for instance, write in another world the epics which the necessity of earning my living may have stifled here, did that other world contain no time, no heroic struggles, or no metrical language. Nor is it clear that my epics, to be perfect, would need to be quite endless. If what is foiled in me is really poetic genius and not simply a tendency toward perpetual motion, it would not help me if in heaven, in lieu of my dreamt-of epics, I were allowed to beget several robust children. In a word, if hereafter I am to be the same man improved I must find myself in the same world corrected. Were I transformed into a cherub or transported into a timeless ecstasy, it is hard to see in what sense I should continue to exist. Those results might be interesting in themselves and might enrich the universe; they would not prolong my life nor retrieve my disasters.

For this reason a future life is after all best represented by those frankly material ideals which most Christians—being Platonists—are wont to despise. It would be genuine happiness for a Jew to rise again in the flesh and live for ever in Ezekiel's New Jerusalem, with its ceremonial glories and civic order. It would be truly agreeable for any man to sit in well-watered gardens with Mohammed, clad in green silks, drinking delicious sherbets, and transfixed by the gazelle-like glance of some young girl, all innocence and fire. Amid such scenes a man might remain himself and might fulfil hopes that he had actually cherished on earth. He might also find his friends again, which in somewhat generous minds is perhaps the thought that chiefly sustains interest in a posthumous existence. But to recognize his friends a man must find them in their bodies, with their familiar habits, voices, and interests; for it is surely an insult to affection to say that he could find them in an eternal formula expressing their idiosyncrasy. When, however, it is clearly seen that another life, to supplement this one, must closely resemble it, does not the magic of immortality altogether vanish? Is such a reduplication of earthly society at all credible? And the prospect of awakening again among houses and trees, among children and dotards, among wars and rumours of wars, still fettered to one personality and one accidental past, still uncertain of the future, is not this prospect wearisome and deeply repulsive? Having passed through these things once and bequeathed them to posterity, is it not time for each soul to rest?

GEORGE SANTAYANA (1863–1952), *The Life of Reason*

From too much love of living,
  From hope and fear set free,
We thank with brief thanksgiving
  Whatever gods may be
That no life lives for ever;
That dead men rise up never;
That even the weariest river
  Winds somewhere safe to sea.

Then star nor sun shall waken,
  Nor any change of light:
Nor sound of waters shaken,
  Nor any sound or sight:
Nor wintry leaves nor vernal,
Nor days nor things diurnal;
Only the sleep eternal
  In an eternal night.

ALGERNON CHARLES SWINBURNE (1837–1909), from 'The Garden of Proserpine'

A man acts according to the desires to which he clings. After death he goes to the next world bearing in his mind the subtle impressions of his deeds; and, after reaping there the harvest of those deeds, he returns again to this world of action. Thus he who has desire continues subject to rebirth.

He who lacks discrimination, whose mind is unsteady and whose heart is impure, never reaches the goal, but is born again and again. But he who has discrimination, whose mind is steady and whose heart is pure, reaches the goal and, having reached it, is born no more.

Upanishads, *c.*500 BC

Your essence was not born and will not die. It is neither being nor nonbeing. It is not a void nor does it have form. It experiences neither pleasure nor pain. If you ponder what it is in you that feels the pain of this sickness, and beyond that you do not think or desire or ask anything, and if your mind dissolves like vapour in the sky, then the path to rebirth is blocked and the moment of instant release has come.

BASSUI (1327–87), Zen Buddhist, comforting a dying person

Take, for instance, all the Chinamen. Which of you here, my friends, sees any fitness in their eternal perpetuation unreduced in numbers? Surely not one of you. At most, you might deem it well to keep a few chosen specimens alive to represent an interesting and peculiar variety of humanity; but as for the rest, what comes in such surpassing numbers, and what you can only imagine in this abstract summary collective manner, must be something of which the units, you are sure, can have no individual preciousness. God himself, you think, can have no use for them. An immortality of every separate specimen must be to him and to the universe as indigestible a load to carry as it is to you. So, engulfing the whole subject in a sort of mental giddiness and nausea, you drift along, first doubting that the mass can be immortal, then losing all assurance in the immortality of your own particular person, precious as you all the while feel and realize the latter to be.

But is not such an attitude due to the veriest lack and dearth of your imagination? You take these swarms of alien kinsmen as they are *for you*: an external picture painted on your retina, representing a crowd oppressive by its vastness and confusion. As they are for you, so you think they positively and absolutely are. *I* feel no call for them, you say; therefore there *is* no call for them. But all the while, beyond this externality which is your way of realizing them, they realize themselves with the acutest internality, with the most violent thrills of life . . . Each of these grotesque or even repulsive aliens is animated by an inner joy of living as hot or hotter than that which you feel beating in your private breast . . . The universe, with every living entity which her resources create, creates at the same time a call for that entity, and an appetite for its continuance—creates it, if nowhere else, at least within the heart of the entity itself. It is absurd to suppose, simply because our private power of sympathetic vibration with other lives gives out so soon, that in the heart of infinite being itself there can be such a thing as plethora, or glut, or supersaturation. It is not as if there were a bounded room where the minds in possession had to move up or make place and crowd together to accommodate new occupants. Each new mind brings its own edition of the universe of space along with it, its own room to inhabit; and these spaces never crowd each other . . .

The heart of being can have no exclusions akin to those which our poor little hearts set up. The inner significance of other lives exceeds all our powers of sympathy and insight. If we feel a significance in our own life which would lead us spontaneously to claim its perpetuity, let us be at least tolerant of like claims made by other lives, however numerous, however unideal they may seem to us to be. Let us at any rate not decide adversely on our own claim, whose grounds we feel directly, because we cannot decide favourably on the alien claims, whose grounds we cannot feel at all. That would be letting blindness lay down the law to sight.

William James (1842–1910), *The Will to Believe*

I do not know whether, officially, there has been any alteration in Christian doctrine . . . But what I do know is that belief in survival after death—the individual survival of John Smith, still conscious of himself as John Smith—is enormously less widespread than it was. Even among professing Christians it is probably decaying: other people, as a rule, don't even entertain the possibility that it might be true . . . I do not want the belief in life after death to return, and in any case it is not likely to return. What I do point out is that its disappearance has left a big hole, and that we ought to take notice of that fact. Reared for thousands of years on the notion that the individual survives, man has got to make a considerable psychological effort to get used to the notion that the individual perishes. He is not likely to salvage civilization unless he can evolve a system of good and evil which is independent of heaven and hell.

George Orwell, 'As I Please', *Tribune*, 1944

Thy nature, immortality! who knows?
And yet who knows it not? It is but life
In stronger thread of brighter colour spun,
And spun for ever . . .
But how great
To mingle interests, converse, amities,
With all the sons of reason, scatter'd wide
Thro' habitable space, wherever born,
Howe'er endow'd! To live free citizens
Cf universal nature! To lay hold
By more than feeble faith on the Supreme!
To call heaven's rich unfathomable mines
(Mines, which support archangels in their state)
Our own! To rise in science, as in bliss,
Initiate in the secrets of the skies!
To read creation; read its mighty plan
In the bare bosom of the Deity!
The plan, and execution, to collate!
To see, before each glance of piercing thought,
All cloud, all shadow, blown remote; and leave
No mystery—but that of love divine . . .

EDWARD YOUNG, *Night Thoughts*

O may I join the choir invisible
Of those immortal dead who live again
In minds made better by their presence: live
In pulses stirred to generosity,
In deeds of daring rectitude, in scorn
For miserable aims that end with self,
In thoughts sublime that pierce the night like stars,
And with their mild persistence urge man's search
To vaster issues . . .
This is the life to come,
Which martyred men have made more glorious
For us who strive to follow. May I reach
That purest heaven, be to other souls
The cup of strength in some great agony,
Enkindle generous ardour, feed pure love,
Beget the smiles that have no cruelty—
Be the sweet presence of a good diffused,
And in diffusion ever more intense.
So shall I join the choir invisible
Whose music is the gladness of the world.

GEORGE ELIOT, from 'O may I join the choir invisible'

Darkling I listen; and for many a time
  I have been half in love with easeful Death,
Call'd him soft names in many a musèd rhyme,
  To take into the air my quiet breath;
Now more than ever seems it rich to die,
  To cease upon the midnight with no pain,
    While thou art pouring forth thy soul abroad
      In such an ecstasy!
  Still wouldst thou sing, and I have ears in vain—
    To thy high requiem become a sod.

Thou wast not born for death, immortal Bird!
  No hungry generations tread thee down;
The voice I hear this passing night was heard
  In ancient days by emperor and clown:
Perhaps the self-same song that found a path
  Through the sad heart of Ruth, when, sick for home,
    She stood in tears amid the alien corn;
      The same that oft-times hath
  Charm'd magic casements, opening on the foam
    Of perilous seats, in faery lands forlorn . . .

JOHN KEATS (1795–1821), from 'Ode to a Nightingale'

Well, all these seamen—sailors and skippers—they
are swallowed forever in their mighty Sea,
gone off on their distant cruises, quite carefree,
they're dead—as sure as they ever got under way.

* * *

—No six-foot hole of earth, no graveyard rats:
they've gone to feed the sharks! The soul of a sailor
doesn't seep up through your potato plots.
  It breathes from every roller.

—See where at the horizon the surges lift
  their amorous bellies and wallow
like a whore in heat, you might say, and half-squiffed . . .
  the lads are there, in the billow's hollow.—

—Listen! the hurricane bawls down the wind! . . .
It's their anniversary.—It quite often returns.—
O poet, keep to yourself your songs of the blind;
—for them: the *De profundis* from the wind's horns.

. . . Naked and green, in the spacious virginal brine,
    may they swirl to infinity!
without shrouds or candles, without nails or pine! . . .
—*Earth-born* parvenus, let them roll in the sea!

TRISTAN CORBIÈRE (1845–75), from 'The End', tr. C. F. MacIntyre

I used to scratch medals, when I was a lad,
On my tin soldiers with a knife.
Except for the two that everyone had,
I got no other honours myself in all my life.

That is not to say that to me it is all the same.
In fact, my *Ideal* is
That after my death (cum grano salis)
A little street should be given my name,
A narrow twisty street with lowdown doors,
Steep stairways and cheap little whores,
Shadows and sloping roof-windows I want.

It would be my haunt.

JOACHIM RINGELNATZ (1883–1934), 'Ambition', tr. Christopher Middleton

—What street is this?
—Mandelstam Street.
—What the hell kind of name is that?
No matter which way you turn it
it comes out crooked.

—He wasn't a straight-edge exactly.
His morals resembled no lily.
And that's why this street (or rather.
to be honest, this sewer)
was given the name
of that Mandelstam.

OSIP MANDELSTAM (1891–1938), tr. Clarence Brown and W. S. Merwin

Man should believe in immortality; he has a right to the belief; it meets the wants of his nature, and he may also believe in the promises of religion. But for a philosopher to attempt to deduce the immortality of the soul from a legend is very weak and ineffectual. For me, the eternal existence of my soul is proved from my idea of activity. If I work on unceasingly till my death, nature is bound to give me another form of being when the present one can no longer sustain my spirit.

GOETHE, *Conversations with Eckermann*, 1829

[*An art critic has said that in Vermeer's* View of Delft *a little patch of yellow wall is 'so well painted that it was, if one looked at it by itself, like some priceless specimen of Chinese art, of a beauty that was sufficient in itself'. Though ordered to rest, the writer Bergotte goes to see the picture.*]

. . . Bergotte ate a few potatoes, left the house, and went to the exhibition. At the first few steps he had to climb, he was overcome by an attack of dizziness. He walked past several pictures and was struck by the aridity and pointlessness of such an artificial kind of art, which was greatly inferior to the sunshine of a windswept Venetian palazzo, or of an ordinary house by the sea. At last he came to the Vermeer which he remembered as more striking, more different from anything else he knew, but in which, thanks to the critic's article, he noticed for the first time some small figures in blue, that the sand was pink, and, finally, the precious substance of the tiny patch of yellow wall. His dizziness increased; he fixed his gaze, like a child upon a yellow butterfly that it wants to catch, on the precious little patch of wall. 'That's how I ought to have written,' he said. 'My last books are too dry, I ought to have gone over them with a few layers of colour, made my language precious in itself, like this little patch of yellow wall.' Meanwhile he was not unconscious of the gravity of his condition. In a celestial pair of scales there appeared to him, weighing down one of the pans, his own life, while the other contained the little patch of wall so beautifully painted in yellow. He felt that he had rashly sacrificed the former for the latter. 'All the same,' he said to himself, 'I shouldn't like to be the headline news of this exhibition for the evening papers.'

He repeated to himself: 'Little patch of yellow wall, with a sloping roof, little patch of yellow wall.' Meanwhile he sank down on to a circular settee; whereupon he suddenly ceased to think that his life was in jeopardy and, reverting to his natural optimism, told himself: 'It's nothing, merely a touch of indigestion from those potatoes, which were under-cooked.' A fresh attack struck him down; he rolled from the settee to the floor, as visitors and attendants came hurrying to his assistance. He was dead. Dead forever? Who can say? Certainly, experiments in spiritualism offer us no more proof than the dogmas of religion that the soul survives death. All that we can say is that everything is arranged in this life as though we entered it carrying a burden of obligations contracted in a former life; there is no reason inherent in the conditions of life on this earth that can make us consider ourselves obliged to do good, to be kind and thoughtful, even to be polite, nor for an atheist artist to consider himself obliged to begin over again a score of times a piece of work the admiration aroused by which will matter little to his worm-eaten body, like the patch of yellow wall painted with so much skill and refinement by an artist destined to be forever unknown and barely identified under the name Vermeer. All these obligations, which have no sanction in our present life, seem to belong to a different world, a world based on kindness, scrupulousness, self-sacrifice, a world entirely different from this one and which we leave in order to be born on this earth, before perhaps returning there to live once again beneath the sway of those unknown laws which we obeyed because we bore their precepts in our hearts, not knowing whose hand had traced them there—those laws to which every profound work of the intellect brings us nearer and which are invisible only—if then!—to fools. So that the idea that Bergotte was not permanently dead is by no means improbable.

They buried him, but all through that night of mourning, in the lighted shop-windows, his books, arranged three by three, kept vigil like angels with outspread wings and seemed, for him who was no more, the symbol of his resurrection.

MARCEL PROUST (1871–1922), *Remembrance of Things Past*, tr. C. K. Scott Moncrieff and Terence Kilmartin

Hereafters

ACCORDING to Sir Thomas Browne, 'a dialogue between two infants in the womb concerning the state of this world, might handsomely illustrate our ignorance of the next.' The warning has not of course discouraged speculation. Heaven has proved less amenable by far to the imagination than the other place: unlike dystopias, utopias are non-places. Even St Paul, Sir Thomas observes, has left only a 'negative description' of it, while that other privileged person, Dante, admits that, such is the radiance of paradise, once you have departed your memory is incapable of retaining it. Words have failed us, too. In 1805 Robert Fellowes explained that the Scriptures decline to set out the pleasures of heaven simply because we should no more comprehend the description 'than a man blind from his birth could be made to understand the precise nature of colours from any words which we could use'.

It seems that heaven is most persuasively intimated through abstractions like 'the flower of Peace, the Rose that cannot wither' and the paradoxes of Giles Fletcher, or the yearning for soft raiment and flowers of such as poor Bessy in *North and South*, Brian Power's trees and golden orioles (a child's idea of Limbo), and the cool repose, quiet and light of the lovely Collect for the Feast of All Souls. That is, if we leave aside the idea that heavenliness lies in observing from a safe distance the miseries of the damned, as is argued here by an Archbishop of Dublin. In 1877 Dean Farrar called this notion 'an abominable fancy', and it is likely to commend many of us more feelingly to the damned than to the blissful voyeurs in the grandstand.

Yet the concept works both ways—the worst torment of the lost consists in witnessing the joy of their opposite numbers—and possesses a force and subtlety absent from the cruder horrors and coarser delights held out both by Christianity and by other faiths. If there is to be a future life, then—in a passage cited in the preceding section—George Santayana recommends one that embodies 'frankly material ideals' or tangible benefits, such as the fish and milk, cakes and ales, of the ancient Egyptians or the 'well provided for' paradise pictured in the passage taken from the Koran. Some of us Anglo-Saxons, at any rate, will have to admit, however, that this all sounds suspiciously like the kind of behaviour that leads to people being sent to hell.

More engaging in that case will be Santayana's idea that in a liberal heaven we might find our friends again—a pleasure orthodoxly (I suppose) but meanly denied by Joseph Hall (yet would the Almighty condone such petty snobbishness in his creation?). In Chateaubriand's opinion purgatory is the most rewarding of the supernatural conditions or departments, for literary people. As he glosses it, it is the most delicate in nuance, though when he finds it superior to heaven and hell in poetry because it offers a future which the other two lack, we are bound to remember that this future lies in one or other of those less poetic places.

Hell is another matter altogether, and more our matter. Chateaubriand is surely correct when he accounts for man's greater skill in depicting hell by relating the pains of its inhabitants to the sufferings of earthly existence. Pains alas are more memorable than pleasures; we are better at tormenting others than at making them happy. Milton's hell we can comprehend, and Burton's hints ('a finger burnt by chance we may not abide'), and the sermon from *A Portrait of the Artist as a Young Man* ('so utterly bound and helpless that, as a blessed saint, saint Anselm, writes in his book on similitudes, they are not even able to remove from the eye a worm that gnaws it'), and Thomas Mann's evocation in *Doctor Faustus*, which (he said later) would have been 'inconceivable without the psychological experience of Gestapo cellars'. Likewise, and even if we have little idea of what that sight would be, we can at least begin to appreciate damnation as expressed by Donne and by Marlowe's Mephistophilis—eternal banishment from the sight of God.

Finally, as far as this anthology goes, and by a variation on Luke 15 : 7, David Hume insists that there should be more lamentation over one man damned than over any number of mere kingdoms overthrown. The sentiment will be applauded by many, including those who find the very notion of damnation laughable, while others will ask whether, in the absence of punishment of any kind, there can possibly be any rewards or recompenses, or anything at all beyond a common indiscriminate oblivion.

Now that Sartre has replaced Dante as our eschatological authority, each statement about the Hereafter becomes more than just a piece of descriptive material about another world. It expresses even more strongly a personal attitude about action in this world.

RAYMOND FIRTH, *The Fate of the Soul*, 1955

O Years! and Age! farewell:
Behold I go
Where I do know
Infinity to dwell.

And these mine eyes shall see
All times, how they
Are lost i' th' sea
Of vast eternity.

Where never moon shall sway
The stars; but she
And night shall be
Drown'd in one endless day.

ROBERT HERRICK (1591–1674), 'Eternity'

My soul, there is a country
Far beyond the stars,
Where stands a wingèd sentry
All skilful in the wars:
There, above noise and danger,
Sweet Peace sits crown'd with smiles,
And One born in a manger
Commands the beauteous files.
He is thy gracious Friend,
And—O my soul, awake!—
Did in pure love descend
To die here for thy sake.
If thou canst get but thither,
There grows the flower of Peace,
The Rose that cannot wither,
Thy fortress, and thy ease.
Leave then thy foolish ranges;
For none can thee secure
But One who never changes—
Thy God, thy life, thy cure.

VAUGHAN, 'Peace'

Dark, deep, and cold the current flows
Unto the sea where no wind blows,
Seeking the land which no one knows.

O'er its sad gloom still comes and goes
The mingled wail of friends and foes,
Borne to the land which no one knows.

Why shrieks for help yon wretch, who goes
With millions, from a world of woes,
Unto the land which no one knows?

Though myriads go with him who goes,
Alone he goes where no wind blows,
Unto the land which no one knows.

For all must go where no wind blows,
And none can go for him who goes;
None, none return whence no one knows.

Yet why should he who shrieking goes
With millions, from a world of woes,
Reunion seek with it or those?

Alone with God, where no wind blows,
And Death, His shadow—doom'd, he goes:
That God is there the shadow shows.

O shoreless Deep, where no wind blows!
And thou, O Land, which no one knows!
That God is All, His shadow shows.

EBENEZER ELLIOTT (1781–1849), 'Plaint'

This ae nighte, this ae nighte,
 —Every nighte and alle,
Fire and sleet and candle-lighte,
 And Christe receive thy saule.

When thou from hence away art pass'd,
 —Every nighte and alle,
To Whinny-muir thou com'st at last;
 And Christe receive thy saule.

If ever thou gavest hosen and shoon,
—Every nighte and alle,
Sit thee down and put them on;
And Christe receive thy saule.

If hosen and shoon thou ne'er gav'st nane,
—Every nighte and alle,
The whinnes sall prick thee to the bare bane;
And Christe receive thy saule.

From Whinny-muir when thou may'st pass,
—Every nighte and alle,
To Brig o' Dread thou com'st at last;
And Christe receive thy saule.

From Brig o' Dread when thou may'st pass,
—Every nighte and alle,
To Purgatory fire thou com'st at last;
And Christe receive thy saule.

If ever thou gavest meat or drink,
—Every nighte and alle,
The fire sall never make thee shrink;
And Christe receive thy saule.

If meat and drink thou ne'er gav'st nane,
—Every nighte and alle,
The fire will burn thee to the bare bane;
And Christe receive thy saule.

This ae nighte, this ae nighte,
—Every nighte and alle,
Fire and sleet and candle-lighte,
And Christe receive thy saule.

ANON., 'A Lyke-Wake Dirge'

The metallic weight of iron;
The glaze of glass;
The inflammability of wood . . .

You will not be cold there;
You will not wish to see your face in a mirror;
There will be no heaviness,
Since you will not be able to lift a finger.

There will be company, but they will not heed you;
Yours will be a journey only of two paces
Into view of the stars again; but you will not make it.

There will be no recognition;
No one, who should see you, will say—
Throughout the uncountable hours—

'Why . . . the last time we met, I brought you some flowers!'

WALTER DE LA MARE (1873–1956), 'De Profundis'

May the ethereal elements not rise up as enemies;
May it come that we shall see the Realm of the Blue Buddha.
May the watery elements not rise up as enemies;
May it come that we shall see the Realm of the White Buddha.
May the earthy elements not rise up as enemies;
May it come that we shall see the Realm of the Yellow Buddha.
May the fiery elements not rise up as enemies;
May it come that we shall see the Realm of the Red Buddha.

May the airy elements not rise up as enemies;
May it come that we shall see the Realm of the Green Buddha.
May the elements of the rainbow colours not rise up as enemies;
May it come that all the Realms of the Buddhas will be seen.
May it come that all the Sounds will be known as one's own sounds;
May it come that all the Radiances will be known as one's own radiances . . .

'Prayer for Guidance', *The Tibetan Book of the Dead* (?8th century), tr. W. Y. Evans-Wentz

Here are cakes for thy body,
Cool water for thy throat,
Sweet breezes for thy nostrils,
And thou art satisfied.

No longer dost thou stumble
Upon thy chosen path,
From thy mind all evil
And darkness fall away.

Here by the river,
Drink and bathe thy limbs,
Or cast thy net, and surely
It shall be filled with fish.

The holy cow of Hapi
Shall give thee of her milk,
The ale of gods triumphant
Shall be thy daily draught.

White linen is thy tunic,
Thy sandals shine with gold;
Victorious thy weapons,
That death come not again.

Now upon the whirlwind
Thou followest thy Prince,
Now thou hast refreshment
Under the leafy tree.

Take wings to climb the zenith,
Or sleep in Fields of Peace;
By day the Sun shall keep thee,
By night the rising Star.

'The Other World', *The Egyptian Book of the Dead*, *c.*3300 BC, tr. Robert Hillyer

'Already' said my host 'You have arrived already?
But by what route, what ingenious *raccourci*?
I half-expected you, it is true,
But I expected someone a little older,
Someone rather less arrogant and impulsive,
Someone a little embittered and despondent,
Someone, in short, not quite *you*.
And now you arrive by some unfair expedient,
Having neglected, no doubt, to pay proper attention to the view:
You arrive a little dazed and flushed,
And you find me hardly ready to receive you, hardly able to cope.
It was inconsiderate of you to die so suddenly,
Placing me in this ridiculous quandary.
I had predicted a great future for you,
A future without happiness or hope;

I had prepared a suitable mausoleum for your reception:
And now you arrive with a bundle of daffodils, a fox-terrier,
And a still unfinished smile.
Really!'

MICHAEL ROBERTS (1902–48), '"Already" said my Host'

Now, the necessary Mansions of our restored selves are those two contrary and incomparable places we call *Heaven* and *Hell*. To define them, or strictly to determine what and where these are, surpasseth my Divinity. That elegant Apostle, which seemed to have a glimpse of Heaven, hath left but a negative description thereof; *which neither eye hath seen, nor ear hath heard, nor can enter into the heart of man*: he was translated out of himself to behold it; but, being returned into himself, could not express it . . . Briefly therefore, where the Soul hath the full measure and complement of happiness; where the boundless appetite of that spirit remains compleatly satisfied, that it can neither desire addition nor alteration; that, I think, is truly Heaven . . . wherever GOD will thus manifest himself, there is Heaven.

I was never afraid of Hell, nor never grew pale at the description of that place. I have so fixed my contemplations on Heaven, that I have almost forgot the Idea of Hell, and am afraid rather to lose the Joys of the one, than endure the misery of the other: to be deprived of them is a perfect Hell, and needs, methinks, no addition to compleat our afflictions . . . they go the fairest way to Heaven that would serve GOD without a Hell; other Mercenaries, that crouch into Him in fear of Hell, though they term themselves the servants, are indeed but the slaves, of the Almighty.

BROWNE, *Religio Medici*

Among the several features which distinguish the Christian hell from Tartarus, one is remarkable above all: the torments experienced by the demons themselves. Pluto, the Judges of the Underworld, the Fates and the Furies, did not suffer along with the guilty. The pains felt by our infernal powers are thus an *additional means* for the imagination, and consequently endow our hell with a *poetic advantage* over the hell of the ancients.

It will at least be allowed that *purgatory* affords Christian poets a genre of *supernatural* unknown to antiquity. There is perhaps nothing more favourable to the Muses than this place of purification, situated on the confines of sorrow and of joy, where the confused sentiments of happiness and of misfortune come together. The gradation of sufferings by reason of past faults, the souls more or less happy, more or less bright, according as they approach more or less closely the twofold eternity of pleasures and of pains, could furnish the pen with touching subjects. Purgatory surpasses heaven and hell in poetry, in that it offers a future which the other two lack.

It is in the nature of man to sympathize only with those things which relate to him, which touch him at some point, such as misfortune, for example. Heaven, where boundless felicity reigns, is too far above the human condition for the soul to be strongly affected by the bliss of the elect: one can interest oneself but little in beings who are perfectly happy. This is why poets have succeeded better in the description of hells: at least humanity is there, and the torments of the guilty remind us of the miseries of our life.

FRANÇOIS RENÉ DE CHATEAUBRIAND (1768–1848), *The Genius of Christianity*

The glory of the great all-mover goes
From end to end of all the world, but in one place
More, in its neighbour less, the radiance glows.
I on the most illuminated floor
Of Paradise have stayed and seen what none
Can call to mind again outside the door.
Because our intellect as it draws near
Depth of desire is made so still and deep,
Memory loses all that once was clear.

DANTE (1265–1321), *Paradiso*, tr. T. W. Ramsey

It is no flaming lustre made of light,
No sweet consent or well-timed harmony,
Ambrosia for to feast the appetite,
Or flowery odour mixed with spicery,
No soft embrace or pleasure bodily;
And yet it is a kind of inward feast,
A harmony that sounds within the breast,
An odour, light, embrace, in which the soul doth rest,

A heavenly feast no hunger can consume,
A light unseen, yet shines in every place,
A sound no time can steal, a sweet perfume
No winds can scatter, an entire embrace
That no satiety can e'er unlace.
  Ingraced into so high a favour there,
  The saints with their beau-peers whole worlds outwear,
And things unseen do see, and things unheard do hear.

GILES FLETCHER (1585–1623), The Celestial City, *Christ's Victory and Triumph*

Hierusalem, my happy home,
  When shall I come to thee?
When shall my sorrows have an end,
  Thy joys when shall I see?

O happy harbour of the saints,
  O sweet and pleasant soil,
In thee no sorrow may be found,
  No grief, no care, no toil . . .

There lust and lucre cannot dwell,
  There envy bears no sway;
There is no hunger, heat, nor cold,
  But pleasure every way . . .

Thy walls are made of precious stones,
  Thy bulwarks, diamonds square;
Thy gates are of right orient pearl,
  Exceeding rich and rare.

Thy turrets and thy pinnacles
  With carbuncles do shine;
Thy very streets are paved with gold
  Surpassing clear and fine.

Thy houses are of ivory,
  Thy windows crystal clear,
Thy tiles are made of beaten gold,
  O God, that I were there.

Within thy gates nothing doth come
  That is not passing clean;
No spider's web, no dirt, no dust,
  No filth may there be seen . . .

Thy gardens and thy gallant walks
  Continually are green;
There grows such sweet and pleasant flowers
  As nowhere else are seen . . .

ANON. (16th century), from 'Hierusalem, my happy home'

Many kinds of rivers flow along in this world system Sukhavati. There are great rivers there, one mile broad, and up to fifty miles broad and twelve miles deep. And all these rivers flow along calmly, their water is fragrant with manifold agreeable odours, in them there are bunches of flowers to which various jewels adhere, and they resound with various sweet sounds. And the sound which issues from these great rivers is as pleasant as that of a musical instrument, which consists of hundreds of thousands of kotis of parts, and which, skilfully played, emits a heavenly music. It is deep, commanding, distinct, clear, pleasant to the ear, touching the heart, delightful, sweet, pleasant, and one never tires of hearing it, it always agrees with one and one likes to hear it, like the words 'Impermanent, peaceful, calm, and not-self'. Such is the sound that reaches the ears of those beings.

And, Ananda, both the banks of those great rivers are lined with variously scented jewel trees, and from them bunches of flowers, leaves, and branches of all kinds hang down. And if those beings wish to indulge in sports full of heavenly delights on those river-banks, then, after they have stepped into the water, the water in each case rises as high as they wish it to—up to the ankles, or the knees, or the hips, or their sides, or their ears. And heavenly delights arise. Again, if beings wish the water to be cold, for them it becomes cold; if they wish it to be hot, for them it becomes hot; if they wish it to be hot and cold, for them it becomes hot and cold, to suit their pleasure. And those rivers flow along, full of water scented with the finest odours, and covered with beautiful flowers, resounding with the sounds of many birds, easy to ford, free from mud, and with golden sand at the bottom. And all the wishes those beings may think of, they all will be fulfilled, as long as they are rightful.

'Description of the Happy Land', Buddhist (1st century AD), tr. Edward Conze

When the sun ceases to shine; when the stars fall down and the mountains are blown away; when camels big with young are left untended and the wild beasts are brought together; when the seas are set alight and men's souls are reunited; when the infant girl, buried alive, is asked for what crime she was slain; when the records of men's deeds are laid open and the heaven is stripped bare; when Hell burns fiercely and Paradise is brought near: then each soul shall know what it has done.

For the unbelievers We have prepared fetters and chains, and a blazing Fire. But the righteous shall drink of a cup tempered at the Camphor Fountain, a gushing spring at which the servants of Allah will refresh themselves . . . Reclining there upon soft couches, they shall feel neither the scorching heat nor the biting cold. Trees will spread their shade around them, and fruits will hang in clusters over them.

When the sky splits asunder and reddens like a rose or stainèd leather (which of your Lord's blessings would you deny?), on that day neither man nor jinnee shall be asked about his sins. Which of your Lord's blessings would you deny?

The wrongdoers shall be known by their looks; they shall be seized by their forelocks and their feet. Which of your Lord's blessings would you deny?

That is the Hell which the sinners deny. They shall wander between fire and water fiercely seething. Which of your Lord's blessings would you deny?

But the true servants of Allah shall be well provided for, feasting on fruit, and honoured in the gardens of delight. Reclining face to face upon soft couches, they shall be served with a goblet filled at a gushing fountain, white, and delicious to those who drink it. It will neither dull their senses nor befuddle them. They shall sit with bashful, dark-eyed virgins, as chaste as the sheltered eggs of ostriches.

The Koran (early 7th century), tr. N. J. Dawood

And I saw a new heaven and a new earth; for the first heaven and the first earth were passed away; and there was no more sea. And I John saw the holy city, new Jerusalem, coming down from God out of heaven, prepared as a bride adorned for her husband. And I heard a great voice out of heaven saying, Behold, the tabernacle of God is with men, and he will dwell with them, and they shall be his people, and God himself shall be with them, and be their God . . .

And the twelve gates were twelve pearls; every several gate was of one pearl: and the street of the city was pure gold, as it were transparent glass. And I saw no temple therein: for the Lord God Almighty and the Lamb are the temple of it. And the city had no need of the sun, neither of the moon, to shine in it: for the glory of God did lighten it, and the Lamb is the light thereof. And the nations of them which are saved shall walk in the light of it: and the kings of the earth do bring their glory and honour into it.

Revelation, 21

'I shall have a spring where I'm boun' to, and flowers, and amaranths, and shining robes besides . . . If yo'd led the life I have, and thought at times, "maybe it'll last for fifty or sixty years—it does wi' some,"—and got dizzy and dazed, and sick, as each of them sixty years seemed to spin about me, and mock me with its length of hours and minutes, and endless bits o' time—oh, wench! I tell thee thou'd been glad enough when th' doctor said he feared thou'd never see another winter.'

'Why, Bessy, what kind of a life has yours been?'

'Nought worse than many others', I reckon. Only I fretted again it, and they didn't . . . I never knew why folk in the Bible cared for soft raiment afore. But it must be nice to go dressed as yo' do. It's different fro' common. Most fine folk tire my eyes out wi' their colours; but somehow yours rest me . . . I used to think once that if I could have a day of doing nothing, to rest me . . . it would maybe set me up. But now I've had many days o' idleness, and I'm just as weary o' them as I was o' my work. Sometimes I'm so tired out I think I cannot enjoy heaven without a piece of rest first. I'm rather afeard o' going straight there without getting a good sleep in the grave to set me up . . . Sometimes, when I've thought o' my life, and the little pleasure I've had in it, I've believed that, maybe, I was one of those doomed to die by the falling of a star from heaven. "And the name of the star is called Wormwood; and the third part of the waters became wormwood; and men died of the waters, because they were made bitter." One can bear pain and sorrow better if one thinks it has been prophesied long before for one: somehow, then it seems as if my pain was needed for the fulfilment; otherways it seems all sent for nothing.'

'Nay, Bessy—think!' said Margaret. 'God does not willingly afflict. Don't dwell so much on the prophecies, but read the clearer parts of the Bible.'

'I dare say it would be wiser; but where would I hear such grand words of promise—hear tell o' anything so far different fro' this dreary world, and this town above a', as in Revelations? Many's the time I've repeated the verses in the seventh chapter to myself, just for the sound. It's as good as an organ, and as different from every day, too. No, I cannot give up Revelations. It gives me more comfort than any other book i' the Bible.'

MRS GASKELL (1810–65), *North and South*

Lord God of mercies, grant to the souls of all thy servants a place of cool repose, the blessedness of quiet, the brightness of light: through our Lord.

Collect, Feast of All Souls

There we shall be with seraphims and cherubims, creatures that will dazzle your eyes to look on them. There, also, we shall meet with thousands and thousands that have gone before us to that place; none of them are hurtful, but loving and holy, every one walking in the sight of God, and standing in his presence with acceptance for ever; in a word, there we shall see the elders with their golden crowns; there we shall see the holy virgins with golden harps; there we shall see men, that by the world were cut in pieces, burnt in flames, eaten of beasts, drowned in the seas, for the love that they bare to the Lord of the place; all well, and clothed with immortality as with a garment.

BUNYAN, *The Pilgrim's Progress*

I find much inquiry of curious wits, whether we shall know one another in Heaven. There is no want of arguments on both parts; and the greatest probabilities have seemed to be for the affirmative. But, O Lord, whether or not we shall know one another, I am sure we shall all, thy glorified Saints, know thee; and in knowing thee we shall be infinitely happy. And what would be more? Surely, as we find here, that the sun puts out the fire, and the greater light ever extinguisheth the less, so why may we not think it to be above? When thou art all in all to us, what can the knowledge of any creature add to our blessedness? And if when we casually meet with a brother or a son before some great prince, we forbear the ceremonies of our mutual respects, as being wholly taken up with the awful regard of a greater presence; how much more may we justly think that, when we meet before the glorious throne of the God of Heaven, all the respects of our former earthly relations must utterly cease, and be swallowed up of that beatifical presence, Divine Love, and infinitely blessed fruition of the Almighty!

JOSEPH HALL (1574–1656), *Susurrium cum Deo*

We have been so long accustomed to the hypothesis of your being taken away from us, especially during the past ten months, that the thought that this may be your last illness conveys no very sudden shock. You are old enough, you've given your message to the world in many ways and will not be forgotten; you are here left alone, and on the other side, let us hope and pray, dear, dear old Mother is waiting for you to join her. If you go, it will not be an inharmonious thing . . . As for the other side, and Mother, and our all possibly meeting, I *can't* say anything. More than ever at this moment do I feel that if that *were* true, all would be solved and justified. And it comes strangely over me in bidding you goodbye how a life is but a day and expresses mainly but a single note. It is so much like the act of bidding an ordinary good night. Good night, my sacred old Father! If I don't see you again—Farewell! a blessed farewell! Your

William

WILLIAM JAMES, to his father, Henry James, Sr, during the latter's final illness in 1882

Though in one respect a view of the misery which the damned undergo might seem to detract from the happiness of the blessed, through pity and commiseration: yet there is another, a nearer, and much more affecting consideration, viz. that all this is the misery which they themselves were often exposed to, and were in imminent danger of incurring; in this view, why may not the sense of their own escape so far overcome the sense of another's ruin, as to extinguish the pain that usually attends the idea of it, and even render it productive of some real happiness?

WILLIAM KING, Archbishop of Dublin, *On the Origin of Evil*, 1702

Lastly I give the poor
woman, my mother, who bore
much pain for me—God knows!
this prayer to our Mistress,
Mary, my house and fortress
against the ills and sorrows
of life. I have no other
patron, nor has my mother.

'Lady of heaven, queen of the world,
and ruler of the underworld,
receive your humble Christian child,
and let him live with those you save;
although my soul is not much worth
saving, my Mistress and My Queen,
your grace is greater than my sin—
without you no man may deserve,
or enter heaven. I do not lie:
in this faith let me live and die.

'Say to your Son that I am his;
Mary of Egypt was absolved,
also the clerk, Theophilus,
whom you consented to restore,
although he'd made a pact with hell.
Save me from ever doing such ill,
our bond with evil is dissolved,
Oh Virgin, undefiled, who bore
Christ whom we celebrate at Mass—
in this faith let me live and die.

'I am a woman—poor, absurd,
who never learned to read your word—
at Mass each Sunday, I have seen
a painted paradise with lutes
and harps, a hell that boils the damned:
one gives me joy, the other doubts.
Oh let me have your joy, my Queen,
bountiful, honest and serene,
by whom no sinner is condemned—
in this faith let me live and die.

'You bore, oh Virgin and Princess,
Jesus, whose Kingdom never ends—
Our Lord took on our littleness,
and walked the world to save his friends—
he gave his lovely youth to death,
that's why I say to my last breath
in this faith let me live and die.'

François Villon (1431–?63), 'Villon's Prayer for his Mother to Say to the Virgin', tr. Robert Lowell

I tried to explain to Y Jieh about confession and the need for a soul to be in a state of grace before communion, but it was very difficult. 'Anyway, Y Jieh,' I said, seeing she was not impressed, 'you and Jieh Jieh won't have to burn in flames when you die, for you will both go to a place called Limbo. That's where all good people go who do not follow the teaching of the Lord of Heaven. In that place you will be able to walk among beautiful trees like the wood in the Russian Park and there will be a river too, and maybe in the spring golden orioles will come and sing to you.'

BRIAN POWER; the author, as a child, is speaking to his amah, in Tientsin, *c.*1925

Some of these heavens are peculiar. One, belonging to a particular lineage, is a Heaven of the Halt, or the Lame; spirits who limp have that as their home. Another is the Heaven of Cannibals, spirits who eat flesh and who have only one nostril, one ear, one leg, one arm, etc., apiece. Other heavens are sloping, others still unstable. I have an account of heavenly doings, telling how a spirit may be standing on a sloping unstable heaven, to the irritation of the owner-spirit. He is annoyed; he wants the other to leave. So he gives his heaven a tilt, and slides the other spirit off, making him take flight to another heaven. He goes off, flying like a bird . . . Unmarried women have their own heaven, as do married women, married men, and bachelors. But the rules of abode in the hereafter are elastic; as in Tikopia, there are fixed dwellings determined by social affiliation and by status in sex and marriage, with additional provision for physical defects. But choice among these dwellings and visiting from one to another is free. The lame spirits and the rest, like the gods, go strolling about as they wish. In the heavens there is eating and drinking, and some spirits go and work in the cultivations. But the great occupation is dancing. In many ways it is a pagan South Seas version of the Elysian fields—or of Marc Connelly's *Green Pastures*.

FIRTH, op. cit.; of the Tikopians, an Oceanic community

You may talk about me just as much as you please,
  Hold the wind, don't let it blow,
I'm gonna talk about you on the bendin' of my knees.

Hold the wind, hold the wind,
Hold the wind, don't let it blow.
Hold the wind, hold the wind,
Hold the wind, don't let it blow.

If you don't believe I have been redeemed,
Just follow me down to the Jordan stream.

My soul got wet in the midnight dew,
And the mornin' star was a witness too.

When I get to Heaven, gwine walk and tell,
Three bright angels go ring them bells.

When I get to Heaven, gwine be at ease,
Me and my God gonna do as we please.

Gonna chatter with the Father, argue with the Son,
Tell um 'bout the world I just come from.

ANON., North American, 'Hold the Wind'

Those—dying then,
Knew where they went—
They went to God's Right Hand—
That Hand is amputated now
And God cannot be found—

The abdication of Belief
Makes the Behaviour small—
Better an ignis fatuus
Than no illume at all—

EMILY DICKINSON, 'Those—dying then'

In paradise the work week is fixed at thirty hours
salaries are higher prices steadily go down
manual labour is not tiring (because of reduced gravity)
chopping wood is no harder than typing
the social system is stable and the rulers are wise
really in paradise one is better off than in whatever country

At first it was to have been different
luminous circles choirs and degrees of abstraction
but they were not able to separate exactly
the soul from the flesh and so it would come here
with a drop of fat a thread of muscle
it was necessary to face the consequences
to mix a grain of the absolute with a grain of clay
one more departure from doctrine the last departure
only John foresaw it: you will be resurrected in the flesh

Not many behold God
He is only for those of 100 per cent pneuma
the rest listen to communiqués about miracles and floods
some day God will be seen by all
when it will happen nobody knows

As it is now every Saturday at noon
sirens sweetly bellow
and from the factories go the heavenly proletarians
awkwardly under their arms they carry their wings like violins

ZBIGNIEW HERBERT (b. 1924), 'Report from Paradise', tr. Czesław Miłosz

I do not believe that Heaven and Hell are in different places,
I do not believe that the utmost anguish of the damned
Could ever damp the bliss of neighbouring Saints.
I do not believe there have ever been complaints
From any of the Twenty Four Elders or Seven Spirits,
About things like the smell of brimstone. 'At first it seems strong'
They confess, 'but one does not notice it for long,
And we keep our incense burning night and day.'
'While for the groaning and gnashing of teeth,' the angels say
'What with the noise of golden harps and new song
They scarcely worry us at all.' To a recent guest
Shy at first amid so much goodness, wondering
Whether one can ever really be friends with the Blest,
Gazing down at the unconsumable Phoenix nest,
At the obstinate host whose daily bread is destruction
Yet none can cease to suffer by being destroyed—
To such, a hospitable Elder will often come
Saying, 'Meet me here when it's dark. You have never enjoyed
Beauty on earth such as I will show you tonight—
The fires of Hell reflected in the Glassy Sea.'

The hours of evening pass; his golden crown, a little tight,
Tires him at first, his unaccustomed wings
Bewilder him, his fingers on the golden strings
Find disconcerting music, and his own voice
Startles him with its raptures when he sings.
Darkness drops; he stands by the smiling Elder
Wing to wing. Shall he look up or down?
At the rose-leaf Phantom caught in the glacier of Heaven?
At the scarlet Fury prancing over Hell-town?
'It's wonderful to look at it, isn't it,' an Elder once said,
'Surely this alone makes it worth while to be dead!'
So Heaven and Hell live side by side
And such troubles as happen are of the mildest kind.
Now and again the dull, the gentle damned
Stir, and some salvaged Lucifer will try
To organize revolt. Which Heaven does not mind.
What does it mean? A few lost spirits clutching
Charred banners with the motto 'We want wings',
Or 'Harps for Hell', or 'Golden crowns for all'.
The unpresentable, scrap-heap Lucifer flings
A written protest over Heaven's wall.
'They're bound to answer', 'This time they must do something—'
The meek spirits whisper, waiting outside.
Hours go by. Suddenly a terrible light
Flashes over them. Is it some new device
For blistering Hell—for cutting off their retreat?
No! That transcendent whiteness is the Angel of Day
Telling them quietly but firmly to go away.

ARTHUR WALEY (1889–1966), 'No Discharge'

No, no, no, my child: do not pray. If you do, you will throw away the main advantage of this place. Written over the gate here are the words 'Leave every hope behind, ye who enter'. Only think what a relief that is! For what is hope? A form of moral responsibility. Here there is no hope, and consequently no duty, no work, nothing to be gained by praying, nothing to be lost by doing what you like. Hell, in short, is a place where you have nothing to do but amuse yourself.

My child: one word of warning first. Let me complete my friend Lucifer's similitude of the classical concert. At every one of those concerts in England you will find rows of weary people who are there, not because they really like classical music, but because they think they ought to like it. Well, there is the same thing in heaven. A number of people sit there in glory, not because they are happy, but because they think they owe it to their position to be in heaven. They are almost all English.

GEORGE BERNARD SHAW (1856–1950), *Man and Superman*

Up to the bed by the window, where I be lyin',
Comes bells and bleat of the flock wi' they two children's clack.
Over, from under the eaves there's the starlings flyin',
And down in yard, fit to burst his chain, yapping out at Sue I do hear
young Mac.

Turning around like a falled-over sack
I can see team ploughin' in Whithy-bush field and meal carts startin'
up road to Church-Town;
Saturday arternoon the men goin' back
And the women from market, trapin' home over the down.

Heavenly Master, I wud like to wake to they same green places
Where I be know'd for breakin' dogs and follerin' sheep.
And if I may not walk in th' old ways and look on th' old faces
I wud sooner sleep.

CHARLOTTE MEW (1869–1928), 'Old Shepherd's Prayer'

The heavenly fields of Paradise,
The happy country, don't tempt me:
I'll find no women in the skies
Lovelier than the ones I see.

No angel with the finest wings
Could substitute there for my wife,
And sitting on a cloud to sing's
Not my choice for the eternal life.

O Lord, I think the best for me's
To leave me in this world, don't you?
But first, heal my infirmities,
And see about some money, too.

The world is full of vice and sin,
I know; I'm used to that, from years
Of pounding the macAdam in
This long and mucky vale of tears.

The hustling world won't get me down—
I hardly ever leave the house:
I like to stay in dressing-gown
And slippers, home, beside the spouse.

Leave me with her! My soul enjoys
So much the music I can hear
In the pure chatter of her voice!
And then her look, so true and clear!

Just good health, and a pay award,
That's all; just living here below
Happily ever after, Lord,
With my wife, in the status quo!

HEINE, *Zum Lazarus*, tr. Alistair Elliot

'Art thou the grave of Charidas?' 'If for Arimmas' son,
The Cyrenean, you inquire, I am the very one.'
'How goes it, Charidas, below?' 'Much gloom.' 'And the way back?'
'A lie, there is none.' 'Pluto, then?' 'Pluto's a myth.' 'Alack!'
'I'm telling you the truth. If you want fairy tales instead,
The market price of oxen here is half a crown a head.'

CALLIMACHUS (*c.*310–*c.*240 BC), *Epigrams*, tr. G. M. Young

F. W. H. Myers, whom spiritualism had converted to belief in a future life, questioned a woman who had lately lost her daughter as to what she supposed had become of her soul. The mother replied: 'Oh well, I suppose she is enjoying eternal bliss, but I wish you wouldn't talk about such unpleasant subjects.'

BERTRAND RUSSELL, *Unpopular Essays*, 1950

O for the time when I shall sleep
Without identity,
And never care how rain may steep
Or snow may cover me!

No promised Heaven, these wild desires
Could all or half fulfil;
No threatened Hell, with quenchless fires,
Subdue this quenchless will!

EMILY BRONTË (1818–48), from 'Enough of Thought, Philosopher'

This world and the other world are like the two wives of one and the same husband: if you please one, you make the other envious.

UNNAMED ARAB SAGE

The lay apologist Caraccioli (1761) was lyrical in his precision and saw Heaven as situated on planets scattered through the universe, with the redeemed, resplendent in their spiritual bodies, sweeping at will between them with the stars as their landmarks . . . The expert on infernal geography in the early part of the century, the Englishman Swinden [*An Enquiry into the Nature and Place of Hell*, 1727], calculated that the accumulation of successive generations had already outrun the sub-terrestrial storage space, and that insufficient air could penetrate to the core of the globe to keep the furnaces hot enough for reasonable efficiency. The sun seemed more appropriate, both in size and temperature, to provide an adequate Hell for the foreseeable future.

JOHN MCMANNERS, *Death and the Enlightenment*, 1981

THROUGH ME YOU ENTER THE CITY OF LAMENT
THROUGH ME YOU ENTER INTO PAIN ETERNAL
THROUGH ME YOU ENTER WHERE THE LOST ARE SENT.
JUSTICE MOVED MY HIGH CREATOR SEMPITERNAL
POWER DIVINE HAS BEEN MY MAKER
WISDOM SUPREME AND PRIMAL LOVE SUPERNAL.
BEFORE ME WERE NO THINGS CREATED, BUT FOR EVER,
AND I FOR ALL ETERNITY ENDURE:
ABANDON EVERY HOPE, ALL YE THAT ENTER.
These words of colour louring and obscure
I saw inscribed on high above a gate . . .

DANTE, *Inferno*, tr. Ronald Bottrall

A dungeon horrible, on all sides round
As one great furnace flamed, yet from those flames
No light, but rather darkness visible
Served only to discover sights of woe,
Regions of sorrow, doleful shades, where peace
And rest can never dwell, hope never comes
That comes to all; but torture without end
Still urges, and a fiery deluge, fed
With ever-burning sulphur unconsumed:
Such place Eternal Justice had prepared
For those rebellious, here their prison ordained
In utter darkness, and their portion set
As far removed from God and light of heav'n
As from the centre thrice to th' utmost pole.

MILTON, *Paradise Lost*

There is a place,
List, daughter! in a black and hollow vault,
Where day is never seen; there shines no sun,
But flaming horror of consuming fires,
A lightless sulphur, chok'd with smoky fogs
Of an infected darkness: in this place
Dwell many thousand thousand sundry sorts
Of never-dying deaths: there damnèd souls
Roar without pity; there are gluttons fed
With toads and adders; there is burning oil
Pour'd down the drunkard's throat; the usurer
Is forc'd to sup whole draughts of molten gold;
There is the murderer forever stabb'd,
Yet can he never die; there lies the wanton
On racks of burning steel, whilst in his soul
He feels the torment of his raging lust.

JOHN FORD (1586–*c.*1640), *'Tis Pity She's a Whore*

The terrible meditation of hell fire and eternal punishment much torments a sinful silly soul. What's a thousand years to eternity? *Ubi moeror*, *ubi fletus*, *ubi dolor sempiternus; mors sine morte*, *finis sine fine*; a finger burnt by chance we may not endure, the pain is so grievous, we may not abide an hour, a night is intolerable; and what shall this unspeakable fire then be that burns for ever, innumerable infinite millions of years, *in omne aevum*, *in aeternum*. O eternity!

BURTON, *The Anatomy of Melancholy*

Faustus. Where are you damn'd?
Mephistophilis. In hell.
Faust. How comes it, then, that thou are out of hell?
Meph. Why, this is hell, nor am I out of it.
Think'st thou that I, who saw the face of God,
And tasted the eternal joys of heaven,
Am not tormented with ten thousand hells
In being depriv'd of everlasting bliss?

CHRISTOPHER MARLOWE (1564–93), *The Tragical History of Doctor Faustus*

What Tophet is not Paradise? What Brimstone is not Amber? What gnashing is not a comfort? What gnawing of the worms is not a tickling? What torment is not a marriage bed, to this damnation to be secluded eternally, eternally, eternally from the sight of God?

DONNE, *Sermons*

The amiable Dr Adams suggested that God was infinitely good. JOHNSON: 'That he is infinitely good, as far as the perfection of his nature will allow, I certainly believe; but it is necessary for good upon the whole, that individuals should be punished. As to an *individual*, therefore, he is not infinitely good; and as I cannot be *sure* that I have fulfilled the conditions on which salvation is granted, I am afraid I may be one of those who shall be damned.' (Looking dismally.) DR ADAMS: 'What do you mean by damned?' JOHNSON (passionately and loudly): 'Sent to hell, Sir, and punished everlastingly.' DR ADAMS: 'I don't believe that doctrine.' JOHNSON: 'Hold, Sir; do you believe that some will be punished at all?' DR ADAMS: 'Being excluded from heaven will be a punishment; yet there may be no great positive suffering.' JOHNSON: 'Well, Sir, but if you admit any degree of punishment, there is an end of your argument for infinite goodness simply considered.' BOSWELL: 'But may not a man attain to such a degree of hope as not to be uneasy from the fear of death?' JOHNSON: 'A man may have such a degree of hope as to keep him quiet. You see I am not quiet, from the vehemence with which I talk; but I do not despair.' MRS ADAMS: 'You seem, Sir, to forget the merits of our Redeemer.' JOHNSON: 'Madam, I do not forget the merits of my Redeemer; but my Redeemer has said that he will set some on his right hand and some on his left.'

BOSWELL, *Life of Johnson*

Hell is the centre of evils and, as you know, things are more intense at their centres than at their remotest points. There are no contraries or admixtures of any kind to temper or soften in the least the pains of hell. Nay, things which are good in themselves become evil in hell. Company, elsewhere a source of comfort to the afflicted, will be there a continual torment: knowledge, so much longed for as the chief good of the intellect, will there be hated worse than ignorance: light, so much coveted by all creatures from the lord of creation down to the humblest plant in the forest, will be loathed intensely. In this life our sorrows are either not very long or not very great because nature either overcomes them by habits or puts an end to them by sinking under their weight. But in hell the torments cannot be overcome by habit, for while they are of terrible intensity they are at the same time of continual variety, each pain, so to speak, taking fire from another and re-endowing that which has enkindled it with a still fiercer flame. Nor can nature escape from these intense and various tortures by succumbing to them for the soul is sustained and maintained in evil so that its suffering may be the greater. Boundless extension of torment, incredible intensity of suffering, unceasing variety of torture—this is what the divine majesty, so outraged by sinners, demands; this is what the holiness of heaven, slighted and set aside for the lustful and low pleasures of the corrupt flesh, requires; this is what the blood of the innocent Lamb of God, shed for the redemption of sinners, trampled upon by the vilest of the vile, insists upon.

—Last and crowning torture of all the tortures of that awful place is the eternity of hell. Eternity! O, dread and dire word. Eternity! What mind of man can understand it? And remember, it is an eternity of pain. Even though the pains of hell were not so terrible as they are, yet they would become infinite, as they are destined to last for ever. But while they are everlasting they are at the same time, as you know, intolerably intense, unbearably extensive. To bear even the sting of an insect for all eternity would be a dreadful torment. What must it be, then, to bear the manifold tortures of hell for ever? For ever! For all eternity! Not for a year or for an age but for ever.

JOYCE, *A Portrait of the Artist as a Young Man*

That is the secret delight and security of hell, that it is not to be informed on, that it is protected from speech, that it just is, but cannot be public in the newspaper, be brought by any word to critical knowledge, wherefor precisely the words 'subterranean', 'cellar', 'thick walls', 'soundlessness', 'forgottenness', 'hopelessness', are the poor, weak symbols. One must just be satisfied with symbolism, my good man, when one is speaking of hell, for there everything ends—not only the word that describes, but everything altogether. This is indeed the chiefest characteristic and what in most general terms is to be uttered about it: both that which the newcomer thither first experiences, and what at first with his as it were sound senses he cannot grasp, and will not understand, because his reason or what limitation soever of his understanding prevents him, in short because it is quite unbelievable enough to make him turn white as a sheet, although it is opened to him at once on greeting, in the most emphatic and concise words, that '*here everything leaves off*'. Every compassion, every grace, every sparing, every last trace of consideration for the incredulous, imploring objection 'that you verily cannot do so unto a soul': it is done, it happens, and indeed without being called to any reckoning in words; in soundless cellar, far down beneath God's hearing, and happens to all eternity.

THOMAS MANN, *Doctor Faustus*, tr. H. T. Lowe-Porter

'There was, they say, here on earth a thinker and philosopher. He rejected everything, "laws, conscience, faith", and, above all, the future life. He died; he expected to go straight to darkness and death and he found a future life before him. He was astounded and indignant. "This is against my principles!" he said. And he was punished for that . . . that is, you must excuse me, I am just repeating what I heard myself, it's only a legend . . . he was sentenced to walk a quadrillion kilometres in the dark (we've adopted the metric system, you know) and when he has finished that quadrillion, the gates of heaven would be opened to him and he'll be forgiven . . .'

'And what tortures have you in the other world besides the quadrillion kilometres?' asked Ivan, with a strange eagerness.

'What tortures? Ah, don't ask. In old days we had all sorts, but now they have taken chiefly to moral punishments—"the stings of conscience" and all that nonsense. We got that, too, from you, from the softening of your manners . . . Reforms, when the ground has not been prepared for them, especially if they are institutions copied from abroad, do nothing but mischief! The ancient fire was better . . .'

'Well, well, what happened when he arrived?'

'Why, the moment the gates of Paradise were open and he walked in, before he had been there two seconds, by his watch (though to my thinking his watch must have long dissolved into its elements on the way), he cried out that those two seconds were worth walking not a quadrillion kilometres but a quadrillion of quadrillions, raised to the quadrillionth power! In fact, he sang "hosannah" and overdid it so, that some persons there of lofty ideas wouldn't shake hands with him at first—he'd become too rapidly reactionary, they said.'

FYODOR MIKHAILOVICH DOSTOEVSKY (1821–81), *The Brothers Karamazov*, tr. Constance Garnett

Judging by pictures
Hell looks more interesting
Than the other place.

Japanese, 18th century

Hell is not interesting; it is merely terrible. Whenever it has not been humanized—as by Dante, who populated it with men of letters and other public figures, thus distracting attention from the penal technicalities—whenever anyone has simply tried to give an original idea of it, even the most imaginative people have not got beyond oafish torments and puerile distortions of earthly peculiarities. But it is precisely the emptiness of the thought of inconceivable, inexorable, everlasting punishment and torment, the premise of a change for the worst impervious to any attempt to reverse it, that has the attraction of an abyss. The same is true of lunatic asylums, which on earth are the ultimate habitation of the lost.

ROBERT MUSIL (1880–1942), *The Man Without Qualities*, tr. Eithne Wilkins and Ernst Kaiser

HELL (BUDDHIST CONCEPTION OF): Eight hells await sinners after death and prior to re-birth, according to Buddhism:

(1) Hell of Repetition (*Samjiva*), for those who kill, lasting 500 years;
(2) Black Rope Hell (*Kāla-Sūtra*), for those who steal, lasting 1000 years;
(3) Crowded Hell (*Samghāta*), for those who abuse sex, lasting 2000 years;
(4) Screaming Hell (*Raurava*), for drunks, lasting 4000 years;
(5) Great Screaming Hell (*Mahā-raurava*), for those who lie, lasting 8000 years;
(6) Hell of Burning Heat (*Tapana*), for those who hold false views (such as not believing in *Tapana*), lasting 16,000 years;
(7) Hell of Great Burning Heat (*Pra-tapana*), for those guilty of sexual defilement of religious people, lasting half a medium *kalpa* (a

kalpa, according to one account, is the length of time it would take an angel who polished a cube-shaped rock measuring eighty leagues just once in every hundred years to completely wear out the rock); and

(8) Hell of No Interval (*Avici*), for murderers, lasting a medium *kalpa*.

St Elmo Nauman, Jr, *Dictionary of Asian Philosophies*, 1978

I waited for His Holiness to get up, but he made no move. There was a silence; only the nose-picking disciple kept up his activities. So I embarked on an anecdote—about the Jesuit priest who was asked how he would reconcile God's all-embracing love with the idea of eternal Hell, and who answered: 'Yes, Hell does exist, but it is always empty.'

I suppose my motive in telling the story was to make him smile again. He did, then said, still smiling: 'We have no eternal Hell in Hinduism; even a little practice of *dharma* will go a long way in accumulating merit.' He quoted a line from the *Gita* in Sanskrit.

Arthur Koestler (b. 1905), *The Lotus and the Robot*: an audience with the Sankaracharya of Kanchi Kamakoti Peetam

Hell is a city much like London—
A populous and a smoky city;
There are all sorts of people undone,
And there is little or no fun done;
Small justice shown, and still less pity.

Shelley, *Peter Bell the Third*

. . . that's what hell will be like, small chat to the babbling of Lethe about the good old days when we wished we were dead.

Samuel Beckett (b. 1906), *Embers*

Falling, I caught the curtain,
Its velvet was the last thing I could feel on earth
As I slid to the floor, howling: aah! aaah!

To the very end I could not believe that I too must . . .
Like everyone.

Then I trod in wheel-ruts
On an ill-paved road. Wooden shacks,
A lame tenement house in a field of weeds.
Potato-patches fenced in with barbed wire.
They played as-if-cards, I smelled as-if-cabbage,
There was as-if-vodka, as-if-dirt, as-if-time.
I said: 'See here . . .', but they shrugged their shoulders,
Or averted their eyes. This land knew nothing of surprise.
Nor of flowers. Dry geraniums in tin cans,
A deception of greenery coated with sticky dust.
Nor of the future. Gramophones played,
Repeating endlessly things which had never been.
Conversations repeated things which had never been.
So that no one should guess where he was, or why.
I saw hungry dogs lengthening and shortening their muzzles,
And changing from mongrel to greyhound, then dachshunds,
As if to signify they were perhaps not quite dogs.
Huge flocks of crows, freezing in mid-air,
Exploded under the clouds . . .

CZESŁAW MIŁOSZ, 'On the Other Side', tr. Jan Darowski (after Swedenborg's various notions of hell)

Eutychides is dead, and what is worse
(fly, wretched shades!) he's coming with his verse.
And listen! they have burned upon his pyre
two tons of music, and a ton of lyre.
You're caught, poor ghosts. But what I want to know
is where in Hell, now he's in hell, to go.

LUCILIUS (*fl.* AD 60), tr. Humbert Wolfe

The lowest circle of hell. Contrary to prevailing opinion it is inhabited neither by despots nor matricides, nor even by those who go after the bodies of others. It is the refuge of artists, full of mirrors, musical instruments, and pictures. At first glance this is the most luxurious infernal department, without tar, fire, or physical tortures.

Throughout the year competitions, festivals, and concerts are held here. There is no climax in the season. The climax is permanent and almost absolute. Every few months new trends come into being and nothing, it appears, is capable of stopping the triumphant march of the avant-garde.

Beelzebub loves art. He boasts that already his choruses, his poets, and his painters are nearly superior to those of heaven. He who has better art has better government—that's clear. Soon they will be able to measure their strength against one another at the Festival of the Two Worlds. And then we will see what remains of Dante, Fra Angelico, and Bach.

Beelzebub supports the arts. He provides his artists with calm, good board, and absolute isolation from hellish life.

ZBIGNIEW HERBERT, 'What Mr Cogito Thinks about Hell', tr. John Carpenter and Bogdana Carpenter

Heaven and hell suppose two distinct species of men, the good and the bad. But the greatest part of mankind float betwixt vice and virtue. Were one to go round the world with an intention of giving a good supper to the righteous and a sound drubbing to the wicked, he would frequently be embarrassed in his choice, and would find, that the merits and demerits of most men and women scarcely amount to the value of either . . . The chief source of moral ideas is the reflection on the interests of human society. Ought these interests, so short, so frivolous, to be guarded by punishments, eternal and infinite? The damnation of one man is an infinitely greater evil in the universe, than the subversion of a thousand millions of kingdoms.

HUME, 'Of the Immortality of the Soul'

Revenants

O THAT it were possible we might
But hold some two days' conference with the dead,

cried the Duchess of Malfi, soon perhaps to have her wish. Two issues ventilated in this section are, Can the spirits of the dead appear to the living? and, If they can, what is their purpose in so doing?

Characteristically, the shortest and sturdiest answer to the first question comes from Dr Johnson: reason says no, something else (here dignified with the rather too precise name of 'belief') says yes. Don Marquis's archy, the cockroach-poet, tells mehitabel the cat that of course he doesn't believe in ghosts:

if you had known
as many of them as i have
you would not
believe in them either . . .

—thus repeating, with a twist, what Mr Flosky says in the excerpt given here from *Nightmare Abbey*, while in turn Flosky is thought to be reproducing Coleridge's curious reply to a question asked at one of his lectures on Shakespeare.

On the second issue much dissatisfaction has been voiced concerning what little, once manifest, these spirits have had to communicate. Either they feed us with trifles or they 'palter with us in a double sense'. In 1900 the President of the Lyons Anthropological Society protested to Camille Flammarion, the French astronomer much occupied with the bearing of psychic phenomena on immortality, that if the dead could return, then all of them would do so, and make themselves useful to their loved ones by saving innocent people from unjust accusations, revealing hidden treasure and so forth—but certainly they wouldn't appear merely to an eccentrically chosen few and then 'talk nonsense to them'.

In the same style are Sir Thomas Browne's amusement over gaps in the knowledge of classical ghosts and J. B. Priestley's complaint about the softening of erstwhile great brains once they have passed over. Explanations have naturally been found for this reticence or taste for trivialities, and Blair conjectures that the laws of their society simply forbid spirits to talk on matters of any real moment. In a fascinating item included here, the spirit of a dead man regrets the obstacles met with in communicating even with those most fitted to understand him—and this despite his having been one of the founders of the Society for Psychical Research.

Hamlet's father obviously suffered from constraints of some kind, but not so the bones of the dead man in the fine lines from the poet Chang Hêng, who was also an astronomer and credited with the invention of the seismograph. Spirits have been forthcoming on occasion, and even helpful. Besides the mysterious voice instrumental in reforming St Augustine, John Aubrey tells of how a neighbouring farmer, indisposed at the time, met in a dream with an old friend, long since deceased, who warned him that if he left his bed he would die. Waking up, the farmer 'rose to make Water, and was immediately seized with a shivering Fit, and died of an Ague, aged 84'. In the course of a seance reported in these pages the spirit of Baudelaire advised a group of young soldiers to study his poems, while the 'affable familiar ghost' of Homer was rumoured to have assisted George Chapman in translating the *Iliad*. Again, the commerce between living and dead in the folklore recorded by Lévi-Strauss proved of benefit to both parties.

The question of vampires and lamias, the Undead, is touched on (not altogether paradoxically) in the section on Love and Death. To confine ourselves here to relations of a gentler nature—though we believe the dead are with us still, and desire them to be, yet their presence may not be invariably welcome. God sees all, but do they need to as well? Tennyson raises this awkward question, and answers it in the only acceptable way: having attained to compassion, understanding and forgiveness, the dead will not expect too much of us.

'Religion', announces Justice Tappercoom in Christopher Fry's play, *The Lady's Not For Burning*,

> Has made an honest woman of the supernatural
> And we won't have it kicking over the traces again,
> Will we, Chaplain?

Religion has done its best. But some believe that sooner or later science will do the job more thoroughly, and reveal the supernatural as having been the common-law wife of the natural all along. Yet the supernatural, I suspect, resembles nature in that, though you drive it out with a pitchfork, it will always come creeping back.

The distance that the dead have gone
Does not at first appear—
Their coming back seems possible
For many an ardent year.

And then, that we have followed them,
We more than half suspect,
So intimate have we become
With their dear retrospect.

EMILY DICKINSON, 'The distance that the dead have gone'

It is wonderful that five thousand years have now elapsed since the creation of the world, and still it is undecided whether or not there has ever been an instance of the spirit of any person appearing after death. All argument is against it; but all belief is for it.

BOSWELL, *Life of Johnson*

The same ignorance makes me so bold as to deny absolutely the truth of the various ghost stories, and yet with the common, though strange, reservation that while I doubt any one of them, still I have faith in the whole of them taken together.

IMMANUEL KANT (1724–1804), *Dreams of a Spirit-Seer*

The introduction of a ghost [in Voltaire's *Semiramis*] was such a bold innovation in a French tragedy, and the dramatist who made this venture justified it with such curious reasons that it is worth dwelling on this for a moment.

'It has been said and written on all sides,' says M. de Voltaire, 'that people no longer believe in ghosts and that apparitions from beyond the grave must seem childish to an enlightened nation.' 'What?' he says in reply. 'When the whole of antiquity believed in these miraculous phenomena, are we not allowed to follow the lead of the ancients? Our religion has sanctified such extraordinary dispensations of providence, so can it be thought ridiculous to revive this belief?' . . .

We no longer believe in ghosts? Who says that? Or rather, what does it mean? Does it mean we are so advanced in our knowledge that we can prove the impossibility of ghosts? Does it mean that certain incontrovertible truths which contradict belief in ghosts have become so widely accepted, are so insistently and invariably present in even the most ordinary person that everything which conflicts with them must of necessity seem to him ridiculous and absurd? No, it can't mean that.

The statement 'we no longer believe in ghosts' can only mean that in this matter, concerning which almost as much can be said for as against, which has not been finally decided and cannot be decided, the prevailing mode of thought has tilted the scales in favour of disbelief. Some people do hold this opinion, and many want to give the impression that they do. These produce all the arguments and set the fashion. The majority keep silent, they express no firm opinion, they cannot make up their minds. During daylight hours they listen with approval when ghosts are ridiculed but in the dead of night they shudder as they listen to tales about them.

GOTTHOLD EPHRAIM LESSING (1729–81), *Hamburgische Dramaturgie*, tr. Idris Parry

Oft in the lone church-yard at night I've seen
By glimpse of moon-shine, chequering thro' the trees,
The schoolboy with his satchel in his hand,
Whistling aloud to bear his courage up,
And lightly tripping o'er the long flat stones
(With nettles skirted, and with moss o'ergrown),
That tell in homely phrase who lie below;
Sudden! he starts, and hears, or thinks he hears
The sound of something purring at his heels:
Full fast he flies, and dares not look behind him,
Till out of breath he overtakes his fellows;
Who gather round, and wonder at the tale
Of horrid *Apparition*, tall and ghastly,
That walks at dead of night, or takes his stand
O'er some new-opened *Grave*; and, strange to tell!
Evanishes at crowing of the cock.

BLAIR, *The Grave*

But the calling back of the dead, or the desirability of calling them back, was a ticklish matter, after all. At bottom, and boldly confessed, the desire does not exist; it is a misapprehension precisely as impossible as the thing itself, as we should soon see if nature once let it happen.

What we call mourning for our dead is perhaps not so much grief at not being able to call them back as it is grief at not being able to want to do so.

THOMAS MANN, *The Magic Mountain*

Tzu-lu asked how one should serve ghosts and spirits. The Master said, Till you have learnt to serve men, how can you serve ghosts? Tzu-lu then ventured upon a question about the dead. The Master said, Till you know about the living, how are you to know about the dead?

The Analects of Confucius (?551–?479 BC), tr. Waley

The departed spirits know things past and to come, yet are ignorant of things present. *Agamemnon* fortells what should happen unto *Ulysses*, yet ignorantly enquires what is become of his own Son. The Ghosts are afraid of swords in *Homer*, yet *Sibylla* tells *Aeneas* in *Virgil*, the thin habit of spirits was beyond the force of weapons. The spirits put off malice with their bodies, and *Caesar* and *Pompey* accord in Latine Hell, yet *Ajax* in *Homer* endures not a conference with *Ulysses*; and *Deiphobus* appears all mangled in *Virgils* Ghosts, yet we meet with perfect shadows among the wounded ghosts of Homer.

BROWNE, *Urn Burial*

Indeed, the spirit world, to which we 'pass on', is more cosy than convincing. The greatest minds that have passed on seem to be deplorably indifferent to our hopes and fears on this side. Surely they could have devised a better method of communication? Even if they had to depend upon trances in darkened rooms, Red Indian 'controls' and the like, surely these deeply curious and formidable intelligences could have contrived to tell us a great deal more about the next world? To ignore us, as they appear to have done, suggests a lamentable change of character. The minds we have praised so often serve us no longer. Moreover, they appear to have done nothing to sharpen the wits of their fellow spirits. There is a disheartening banality about this life beyond the veil. All are happy over there, we are often told, but some of us feel we would find it hard to share that happiness, unless our minds are to be so softened and sweetened they would be like marshmallows. It has been argued that these reports from the other side are necessarily shaped and coloured by the personalities of the mediums—and their 'controls'. So if you attend a seance in Blackpool, let us say, you will probably be given a Blackpool view of the next world. Fair enough! But this returns us brutally to the question of communication. Why these dubious methods—and no other? Why have no better arrangements been devised for this progress of the immortal soul?

However I don't mean to suggest we are doing nothing here but grubbing about on a rubbish heap . . .

J. B. PRIESTLEY (b. 1894), *Over the Long High Wall*

Knock, knock, knock!

'Dear spirit! Is that really you, Napoleon?'

'Yes. What do you wish?'

'It would be so good of you if you'd go and find the Virgin Mary for us, for we want to ask her for some information about the apparitions of Lourdes.'

'All right, my friends. Wait a minute.'

Knock, knock, knock!

'Is this the Virgin Mary?'

'No, she's busy. But here's Messalina.'

CAMILLE FLAMMARION, *Death and Its Mystery*, Vol. III, 1922, tr. Latrobe Carroll (on false notions of 'spiritism')

Anno 1670, not far from Cyrencester, was an Apparition: Being demanded, whether a good Spirit, or a bad? returned no answer, but disappeared with a curious Perfume and most melodious Twang. Mr W. Lilley believes it was a Fairie.

JOHN AUBREY (1626–97), *Miscellanies*

Tell us! ye dead! Will none of you in pity
To those you left behind disclose the secret?
Oh! that some courteous ghost would blab it out!
What 'tis you are, and we must shortly be.
I've heard, that souls departed have sometimes
Forewarn'd men of their death: 'twas kindly done
To knock, and give th' alarum. But what means
This stinted charity? 'Tis but lame kindness
That does its work by halves. Why might you not
Tell us what 'tis *to die*? Do the strict laws

Of your society forbid your speaking
Upon a point so nice? I'll ask no more;
Sullen, like lamps in sepulchres, your shine
Enlightens but yourselves. Well,—'tis no matter;
A very little time will clear up all,
And make us learn'd as you are, and as close.

BLAIR, *The Grave*

The nearest simile I can find to express the difficulties of sending a message—is that I appear to be standing behind a sheet of frosted glass—which blurs sight and deadens sounds—dictating feebly—to a reluctant and somewhat obtuse secretary. A feeling of terrible impotence burdens me—I am so powerless to tell what means so much—I cannot get into communication with those who would understand and believe me.

F. W. H. MYERS, through the automatic writing of a medium in 1906, five years after his death: *Proceedings*, Society for Psychical Research, Vol. xxi

The Rev. Mr Larynx. We have such high authority for ghosts, that it is rank scepticism to disbelieve them. Job saw a ghost, which came for the express purpose of asking a question, and did not wait for an answer.

The Hon. Mr Listless. Because Job was too frightened to give one.

The Rev. Mr Larynx. Spectres appeared to the Egyptians during the darkness with which Moses covered Egypt. The witch of Endor raised the ghost of Samuel. Moses and Elias appeared on Mount Tabor. An evil spirit was sent into the army of Sennacherib, and exterminated it in a single night.

Mr Toobad. Saying, The devil is come among you, having great wrath.

Mr Flosky. Saint Macarius interrogated a skull, which was found in the desert, and made it relate, in presence of several witnesses, what was going forward in hell. Saint Martin of Tours, being jealous of a pretended martyr, who was the rival saint of his neighbourhood, called up his ghost, and made him confess that he was damned. Saint Germain, being on his travels, turned out of an inn a large party of ghosts, who had every night taken possession of the *table d'hôte*, and consumed a copious supper.

Mr Hilary. Jolly ghosts, and no doubt all friars. A similar party took possession of the cellar of M. Swebach, the painter, in Paris, drank his wine, and threw the empty bottles at his head.

The Rev. Mr Larynx. An atrocious act.

Mr Flosky. Pausanius relates that the neighing of horses and the tumult of combatants were heard every night on the field of Marathon: that those who went purposely to hear these sounds suffered severely for their curiosity; but those who heard them by accident passed with impunity.

The Rev. Mr Larynx. I once saw a ghost myself, in my study, which is the last place where anyone but a ghost would look for me. I had not been into it for three months, and was going to consult Tillotson, when, on opening the door, I saw a venerable figure in a flannel dressing-gown, sitting in my armchair, and reading my Jeremy Taylor. It vanished in a moment, and so did I; and what it was or what it wanted I have never been able to ascertain . . .

Mr Flosky. I can safely say I have seen too many ghosts myself to believe in their external existence. I have seen all kinds of ghosts: black spirits and white, red spirits and grey. Some in the shapes of venerable old men, who have met me in my rambles at noon; some of beautiful young women, who have peeped through my curtains at midnight.

The Hon. Mr Listless. And have proved, I doubt not, 'palpable to feeling as to sight'.

Mr Flosky. By no means, sir. You reflect upon my purity. Myself and my friends, particularly my friend Mr Sackbut, are famous for our purity. I see a ghost at this moment . . .

THOMAS LOVE PEACOCK (1785–1866), *Nightmare Abbey*

One evening we gathered in my room, about a small, round, three-legged table. We had placed this table in the very centre of the room, with only our four chairs around it; all the other furniture had been moved away. We examined everything, so that we could see that there could be no tricks, and that no strings were tied to anything. On the mantelpiece were two lighted lamps.

We promised one another that we would do nothing either to help or to hinder anything that might take place, and sat down, with our hands flat on the table, forming a continuous chain with our fingers.

Ten minutes passed without anything happening. We were serious, and in a rather painful state, perhaps (at least I was), but were not in the least nervous. I was praying, under my breath: 'If there is really something beyond terrestrial life, may a gleam come to us from this unknown source of light.'

Suddenly, within the table—in the wood of the table, seemingly—a sharp blow was struck. We looked at one another. This cracking noise seemed to me so characteristic, of such a special kind, that the idea that it might have been caused by one of my three friends did not occur to me, and I felt a shiver run through me from head to foot.

Soon another sharp blow was struck; the table rose on two of its legs and struck three very distinct blows. I had the feeling that the cracking noise could not have been caused by any of us, but that the movement of the table, in striking the floor with one of its legs, might have been so caused, and without a doubt we all had the same thought: that perhaps without wishing to, one or the other of us, bearing down too hard, had pulled the table toward him.

We confided these thoughts to one another, honestly, and then decided to make use of the alphabet, and agreed that the various letters should be designated by the number of blows. After stipulating, besides this, that one blow should mean 'no' and two blows should mean 'yes', we sat down again.

It was not long before the table tilted again. I asked:

'Is this table being moved?'

'Yes.'

'May I know who is moving it?'

'Spirit.'

'Spirit? The spirit of whom?—of one of us?'

'No.'

'Have you a name?'

'Yes; Baudelaire.'

The blows had been struck distinctly, and the letters designated without any mistake. One of the party, even if we had not been watching him, could not have made the table rap with such precision. In a painful state, we looked at one another, without daring to say anything. The table answered some questions as to the existence of the soul after death, and as to certain great moral and religious subjects; it stated the dominant defect of each one of us, and advised: 'Read "Fleurs du Mal".'

FLAMMARION, op. cit.

The Japan Red Cross evinced it would secede from the Executive Committee for the Consolation of Souls of Dead Chinese POWs, denouncing the latter for being too Leftist-inclined.

Japanese news item, 1953

Inscription on a wall in Bermondsey Antique Market:
John Wayne is dead
Underneath, in another and more ghostly hand:
The hell I am

Evening Standard, 1980

'I wonder what these ghosts of mail coaches carry in their bags,' said the landlord, who had listened to the whole story with profound attention.

'The dead letters, of course,' said the Bagman.

'Oh, ah! To be sure,' rejoined the landlord. 'I never thought of that.'

DICKENS, *The Pickwick Papers*

Squats on a toad-stool under a tree
A bodiless childfull of life in the gloom,
Crying with frog voice, 'What shall I be?
Poor unborn ghost, for my mother killed me
Scarcely alive in her wicked womb.
What shall I be? shall I creep to the egg
That's cracking asunder yonder by Nile,
And with eighteen toes,
And a snuff-taking nose,
Make an Egyptian crocodile?' . . .

THOMAS LOVELL BEDDOES (1803–49), *Death's Jest Book*

What beck'ning ghost, along the moonlight shade
Invites my step, and points to yonder glade?
'Tis she!—but why that bleeding bosom gor'd,
Why dimly gleams the visionary sword?
Oh ever beauteous, ever friendly! tell,
Is it, in heav'n, a crime to love too well?
To bear too tender, or too firm a heart,
To act a Lover's or a *Roman*'s part?
Is there no bright reversion in the sky,
For those who greatly think, or bravely die?

POPE, from 'Elegy to the Memory of an Unfortunate Lady'

These flowers are I, poor Fanny Hurd,
Sir or Madam,
A little girl here sepultured.
Once I flit-fluttered like a bird
Above the grass, as now I wave
In daisy shapes above my grave,
All day cheerily,
All night eerily!

—I am one Bachelor Bowring, 'Gent',
Sir or Madam;
In shingled oak my bones were pent;
Hence more than a hundred years I spent
In my feat of change from a coffin-thrall
To a dancer in green as leaves on a wall,
All day cheerily,
All night eerily!

—I, these berries of juice and gloss,
Sir or Madam,
Am clean forgotten as Thomas Voss;
Thin-urned, I have burrowed away from the moss
That covers my sod, and have entered this yew,
And turned to clusters ruddy of view,
All day cheerily,
All night eerily!

—The Lady Gertrude, proud, high-bred,
Sir or Madam,
Am I—this laurel that shades your head;
Into its veins I have stilly sped,
And made them of me; and my leaves now shine,
As did my satins superfine,
All day cheerily,
All night eerily!

—I, who as innocent withwind climb,
Sir or Madam,
Am one Eve Greensleeves, in olden time
Kissed by men from many a clime,
Beneath sun, stars, in blaze, in breeze,
As now by glowworms and by bees,
All day cheerily,
All night eerily!

—I'm old Squire Audeley Grey, who grew,
Sir or Madam,
Aweary of life, and in scorn withdrew;
Till anon I clambered up anew
As ivy-green, when my ache was stayed,
And in that attire I have longtime gayed
All day cheerily,
All night eerily!

—And so these maskers breathe to each
Sir or Madam
Who lingers there, and their lively speech
Affords an interpreter much to teach,
As their murmurous accents seem to come
Thence hitheraround in a radiant hum,
All day cheerily,
All night eerily!

HARDY, 'Voices from Things Growing in a Churchyard'

I am thy father's spirit,
Doom'd for a certain term to walk the night;
And for the day confin'd to fast in fires,
Till the foul crimes done in my days of nature
Are burnt and purg'd away. But that I am forbid
To tell the secrets of my prison-house,
I could a tale unfold, whose lightest word
Would harrow up thy soul, freeze thy young blood,
Make thy two eyes, like stars, start from their spheres,
Thy knotted and combined locks to part,
And each particular hair to stand an end,
Like quills upon the fretful porpentine:
But this eternal blazon must not be
To ears of flesh and blood . . .

SHAKESPEARE, *Hamlet*

I believe . . . that those apparitions and ghosts of departed persons are not the wandring souls of men, but the unquiet walks of Devils, prompting and suggesting us unto mischief, blood, and villainy; instilling and stealing into our hearts that the blessed Spirits are not at rest in their graves, but wander solicitous of the affairs of the World. But that those phantasms appear often, and do frequent Cemeteries, Charnel-houses, and Churches, it is because those are the dormitories of the dead, where the Devil, like an insolent Champion, beholds with pride the spoils and Trophies of his Victory over Adam.

BROWNE, *Religio Medici*

When by thy scorn, O murd'ress, I am dead,
And that thou thinkst thee free
From all solicitation from me,
Then shall my ghost come to thy bed,
And thee, feign'd vestal, in worse arms shall see;
Then thy sick taper will begin to wink,
And he, whose thou art then, being tir'd before,
Will, if thou stir, or pinch to wake him, think
Thou call'st for more,
And in false sleep will from thee shrink;
And then, poor aspen wretch, neglected thou,
Bath'd in a cold quicksilver sweat, wilt lie,
A verier ghost than I:
What I will say, I will not tell thee now,
Lest that preserve thee; and since my love is spent,
I had rather thou should'st painfully repent,
Than by my threat'nings rest still innocent.

DONNE, 'The Apparition'

Have you seen walking through the village
A man with downcast eyes and haggard face?
That is my husband who, by secret cruelty
Never to be told, robbed me of my youth and my beauty;
Till at last, wrinkled and with yellow teeth,
And with broken pride and shameful humility,
I sank into the grave.
But what think you gnaws at my husband's heart?
The face of what I was, the face of what he made me!
These are driving him to the place where I lie.
In death, therefore, I am avenged.

EDGAR LEE MASTERS, 'Ollie McGee'

Nobody will sing a mass,
And no kaddish will be said,
Nothing said and nothing sung
In the first days I am dead.

But perhaps some later day
When the weather's mild and clean,
Frau Mathilde will go walking
On Montmartre with Pauline.

With a crown of everlastings
Come to decorate and sigh
Pauvre homme! to my grave,
Sadness welling in her eye.

Pity, I live too high up
And I can't produce a seat
For my darling here. Oh, she's
Tottering on her tired feet.

Sweet, fat child, no, no, you mustn't
Think of walking home. Ah, wait:
There's a cab-rank—can you see?—
At the cemetery gate.

HEINE, 'Anniversary', tr. Alistair Elliot

Dear little Mosca,
so they called you, I don't know why,
this evening almost in the dark,
while I was reading Deutero-Isaiah
you reappeared beside me,
but without your glasses,
so that you could not see me,
nor could I recognize you in the haze
without that glitter.

Poor Mosca without glasses or antennae,
who had wings only in imagination,
a worn-out and dismantled Bible,
and not very dependable either,
night's black, a flash, a peal of thunder
and then not even the storm.

Could you have left so soon
even without talking?
But it's ridiculous to think
you still had lips.

MONTALE, *Xenia I*, tr. G. Singh

I said, several notebooks ago, that even if I got what seemed like an assurance of H.'s presence, I wouldn't believe it. Easier said than done. Even now, though, I won't treat anything of that sort as evidence. It's the *quality* of last night's experience—not what it proves but what it was—that makes it worth putting down. It was quite incredibly unemotional. Just the impression of her *mind* momentarily facing my own. Mind, not 'soul' as we tend to think of soul. Certainly the reverse of what is called 'soulful'. Not at all like a rapturous re-union of lovers. Much more like getting a telephone call or a wire from her about some practical arrangement. Not that there was any 'message'—just intelligence and attention. No sense of joy or sorrow. No love even, in our ordinary sense. No un-love. I had never in any mood imagined the dead as being so—well, so business-like. Yet there was an extreme and cheerful intimacy. An intimacy that had not passed through the senses or the emotions at all.

C. S. LEWIS, *A Grief Observed*

During those thirty minutes, he held the pencil over the sheet of paper and it moved and filled the pages in large letters . . . At one stage, the pencil said, 'The lady is here, but will not communicate with her husband directly yet. By and by, perhaps, when she is calmer. She is somewhat agitated today, since this is her first effort to communicate with her husband. She is disturbed by the grief of her husband . . .'

In course of time, my wife was able to communicate directly at Mr Rao's sittings. Week after week, she gave me lessons on how to prepare myself so as to be able to communicate my thoughts or receive hers without an intermediary. At the thirty-minute sitting, she criticized my performance in the preceding week. 'It is no use, your sitting up with such rigid concentration: that's just what I do not want. I want you to relax your mind; try to make your mind passive; you can think of me without desperation and also make your mind passive . . . no, no, it's not the rigour of a yogi's meditation that I suggest; this is a more difficult thing, create a channel of communication and wait. Keep your mind inactive . . . I can see that you still worry too much about the child . . . Take good care of her, but don't cramp her with so much anxious thought, which has grown into a habit with you . . . Two nights ago, when you were about to fall asleep, your mind once again wandered off to the sick-bed scenes and the day you mourned my passing over . . . No harm in your remembering those times, but at the root there is still a rawness and that interferes with your perception. Until you can think of me without pain, you will not succeed in your attempts. Train your mind properly and you will know that I am at your side. Not more than ten minutes at a time should you continue the attempt; longer than that, it is likely to harm your health . . . Take care of yourself . . . I am watching the child, and often times she knows I'm there, but she won't talk to you about it . . . She may sometimes take it to be a dream . . . For instance, the other night, you remember a wedding procession that passed down your road, you were all at the gate to watch it, leaving her asleep in your room . . . I approached her at that moment; if you had ever questioned her next morning, what she dreamt, she would have told you point blank, "I dreamt of Raji . . ." Sometimes she may not remember, often she will not care to talk . . . Children are much more cautious than you think . . . Children are precociously cautious. After coming over, I have learnt so much more about the human mind, whose working I can directly perceive . . . In your plane, your handicap is the density of the matter in which you are encased. Here we exist in a more refined state, in a different medium . . . I wish I could explain all that I see, think, and feel . . . When you are prepared for it, I'll be able to tell you much . . .'

R. K. NARAYAN (b. 1907), *My Days*

Oh would that I were a reliable spirit careering around
Congenially employed and no longer by *feebleness* bound
Oh who would not leave the flesh to become a reliable spirit
Possibly travelling far and acquiring merit.

STEVIE SMITH, 'Longing for Death because of Feebleness'

St Augustin heard a Voice, *Tolle, lege*. He took up his Bible, and dipt on *Rom.* 13, 13, *Not in rioting and drunkenness, not in chambering and wantonness*, &c., and reformed his Manners upon it.

AUBREY, op. cit.

Some societies let their dead rest; provided homage is paid to them periodically, the departed refrain from troubling the living. If they come back, they do so only at intervals and on specified occasions. And their return is salutary, since through their influence the dead ensure the regular return of the seasons, and the fertility of gardens and women. It is as if a contract had been concluded between the dead and the living: in return for being treated with a reasonable degree of respect, the dead remain in their own abode, and the temporary meetings between the two groups are always governed by concern for the interests of the living. There is a universal theme in folklore which expresses this formula very clearly: the theme of the *grateful corpse*. A rich hero buys a corpse from creditors who were objecting to its being buried, and has it interred. The dead man appears to his benefactor in a dream and promises him success, on condition that the advantages gained are shared fairly between the two of them. The hero soon wins the love of a princess, whom he rescues from many dangers with the help of his supernatural protector. The question now arises: must the princess be shared? She, as it happens, is bewitched, half woman, half dragon or snake. The dead man claims his share; the hero agrees and his partner, pleased by his fairness, takes only the diabolical part, leaving the hero a humanized wife.

CLAUDE LÉVI-STRAUSS (b. 1908), *Tristes Tropiques*, tr. John and Doreen Weightman

Why do we return? Not in the darkened rooms
Of rattling tambourines and butter muslin;
But as you boil an egg or make the bed
You hear us and answer 'Darling?'

Yes, that's our wish, after all, whatever ancient
Boredom or intervening cause of unwelcome
Would face us, for our presence once again
To be taken all for granted.

We don't come in actuality, alas!
For we're in a place that even cosmologists,
Speculating on collapsed stars and anti-matter,
Couldn't find more alien.

ROY FULLER (b. 1912), from 'Ghost Voice'

Tugging my forelock fathoming Xenophon
grimed Greek exams with grease and lost me marks,
so I whisper when the barber asks *Owt on?*
No, thank you! YES! Dad's voice behind me barks.

They made me use dad's hair-oil to look 'smart'.
A parting scored the grease like some slash scar.
Such aspirations hair might have for ART
were lopped, and licked by dollops from his jar.

And if the page I'm writing on has smears
they're not the sort to lose me marks for mess
being self-examination's grudging tears
soaked into the blotter, Nothingness,
on seeing the first still I'd ever seen
of Rudolph Valentino, father, O
now, *now* I know why you used *Brilliantine*
to slick back your black hair so long ago.

TONY HARRISON (b. 1937), 'Still'

. . . at that time the Pest was in London, he being in the country at Sir Robert Cotton's house with old Camden, he saw in a vision his eldest son (then a child and at London) appear unto him with the mark of a bloody cross on his forehead as if it had been cut with a sword, at which amazed he prayed unto God, and in the morning he came to Mr Camden's chamber to tell him, who persuaded him it was but an apprehension of his fantasy at which he should not be disjected. In the meantime come there letters from his wife of the death of that boy in the plague. He appeared to him, he said, of a manly shape and of that growth that he thinks he shall be at the resurrection.

DRUMMOND OF HAWTHORNDEN, *Ben Jonson's Conversations with Drummond*

When I was nearing forty the person I loved most in the world died, far from me. My mind would have accompanied her anywhere rather than to her death. I should have known and did not even guess. Toward noon I was strolling in a denuded garden, the only one on this South Tyrolean mountain. All was still; then I was called: from the house, I thought. So little prepared was I, that in the first moment it did not occur to me: No one here calls me by my given name. Later in the day came the telegram with the news.

HEINRICH MANN, on the suicide of his sister Carla in 1910; extract from *Ein Zeitalter wird besichtigt*, tr. Richard Winston

We were lying off Victoria [Cameroons]. I had gone down to my cabin thinking to write some letters. I drew aside the door curtain and stepped inside and to my amazement I saw Wilfred sitting in my chair. I felt shock run through me with appalling force and with it I could feel the blood draining away from my face. I did not rush towards him but walked jerkily into the cabin—all my limbs stiff and slow to respond. I did not sit down but looking at him I spoke quietly: 'Wilfred, how did you get here?' He did not rise and I saw that he was involuntarily immobile, but his eyes which had never left mine were alive with the familiar look of trying to make me understand; when I spoke his whole face broke into his sweetest and most endearing dark smile. I felt no fear—I had not when I first drew my door curtain and saw him there; only exquisite mental pleasure at thus beholding him . . . I spoke again. 'Wilfred dear, how can you be here, it's just not possible . . .' But still he did not speak but only smiled his most gentle smile. This not speaking did not now as it had done at first seem strange or even unnatural; it was not only in some inexplicable way perfectly natural but radiated a quality which made his presence with me undeniably right and in no way out of the ordinary. I loved having him there: I could not, and did not want to try to understand how he had got there . . . I could not question anything, the meeting in itself was complete and strangely perfect. He was in uniform and I remember thinking how out of place the khaki looked among the cabin furnishings. With this thought I must have turned my eyes away from him; when I looked back my cabin chair was empty . . .

I felt the blood run slowly back to my face and looseness into my limbs and with these an overpowering sense of emptiness and absolute loss . . . I wondered if I had been dreaming but looking down I saw that I was still standing. Suddenly I felt terribly tired and moving to my bunk I lay down; instantly I went into a deep oblivious sleep. When I woke up I knew with absolute certainty that Wilfred was dead.

HAROLD OWEN (1897–1971), *Journey from Obscurity*, Vol. III (Wilfred, Harold's brother, had in fact been killed on the Western Front several weeks earlier)

It is near Toussaints, the living and dead will say:
'Have they ended it? What has happened to Gurney?'
And along the leaf-strewed roads of France many brown shades
Will go, recalling singing, and a comrade for whom also they
Had hoped well. His honour them had happier made.
Curse all that hates good. When I spoke of my breaking
(Not understood) in London, they imagined of the taking

Vengeance, and seeing things were different in future.
(A musician was a cheap, honourable and nice creature.)
Kept sympathetic silence; heard their packs creaking
And burst into song—Hilaire Belloc was all our master.
On the night of all the dead, they will remember me,
Pray Michael, Nicholas, Maries lost in Novembery
River-mist in the old City of our dear love, and batter
At doors about the farms crying 'Our war poet is lost.
Madame—no bon!'—and cry his two names, warningly, sombrely.

IVOR GURNEY (1890–1937), 'It Is Near Toussaints'

At those who come to my grave with flowers, I can but laugh,
Those, ignorant, unheeding, what you please,
Who think I bear some relation to this stone
And do not know I am in those flowers, and these.

CAHIT SITKI TARANCI (1910–56), 'One of the Dead Speaks', tr. Nermin Menemencioğlu

The dead man answered me:
'In death I rest and am at peace; in life I toiled and strove.
Is the hardness of the winter stream
Better than the melting of spring?
All pride that the body knew,
Was it not lighter than dust?...
Of the Primal Spirit is my substance; I am a wave
In the river of Darkness and Light.
The Maker of All Things is my Father and Mother,
Heaven is my bed and earth my cushion,
The thunder and lightning are my drum and fan,
The sun and moon my candle and my torch,
The Milky Way my moat, the stars my jewels.
With Nature am I conjoined;
I have no passion, no desire.
Wash me and I shall be no whiter,
Foul me and I shall yet be clean.
I come not, yet am here;
Hasten not, yet am swift.'
The voice stopped, there was silence.

CHANG HÊNG (78–139), from 'The Bones of Chuang Tzu', tr. Waley

Do we indeed desire the dead
  Should still be near us at our side?
  Is there no baseness we would hide?
No inner vileness that we dread?

Shall he for whose applause I strove,
  I had such reverence for his blame,
  See with clear eye some hidden shame
And I be lessened in his love?

I wrong the grave with fears untrue:
  Shall love be blamed for want of faith?
  There must be wisdom with great Death:
The dead shall look me through and through.

Be near us when we climb or fall:
  Ye watch, like God, the rolling hours
  With larger other eyes than ours,
To make allowance for us all.

TENNYSON, *In Memoriam A.H.H.*

Handle a large kingdom with as gentle a touch
  as if you were cooking small fish.
If you manage people by letting them alone,
Ghosts of the dead shall not haunt you.
Not that there are no ghosts
But that their influence becomes propitious
In the sound existence of a living man:
There is no difference between the quick and the dead,
They are one channel of vitality.

LAO TZU (6th century BC), *Tao te ching*, tr. Witter Bynner

War, Plague and Persecution

HAMLET remarked of the Norwegian expedition that the 'patch of ground' they were fighting over was 'not tomb enough and continent to hide the slain'. Of a casualty in France during the First World War Wilfred Owen wrote: 'Was it for this the clay grew tall?' Then, 'life goes on with knitting with war with business,' a French voice is heard to say; and whatever the gains and the losses elsewhere, 'What difference does it make to the dead, the orphans and the homeless,' Gandhi asks, 'whether the mad destruction is wrought under the name of totalitarianism or the holy name of liberty or democracy?' The only emperor whose victory is unequivocal is Death; and 'life goes on with the cemetery'. In such conditions the privilege Rilke described as 'one's own death' is not always granted. Every bullet may have its billet, as William of Orange claimed, but so far no one has talked about a nuclear missile having somebody's name on it. Today the old soldier Švejk would reckon that people are much more than twice as clever as they were before.

We know about the horrors of war, and when the time comes we are able to forget them, it seems. It was not my intention here to make an anthology of war writing, and I have selected sparingly from verse and prose dating from the fifth century BC onwards. It is forbidden to take life, Voltaire observed, and hence all who do so are punished, unless they kill in large companies and to the sound of trumpets. I have not attempted to compete with press and television in the matter of civil slaughter. As for natural (as distinct from man-made) disasters and what is termed accidental death—plague ('that was the time I came into my own') must stand in for them all: earthquakes, landslides, storms at sea and on land, fire, flood, drought and famine . . .

Persecution, in whatever shape and whether deemed a natural or a man-made phenomenon, could hardly go unrepresented, it having proved a notable dealer of death in our times. In this category, particularly striking is Nadezhda Mandelstam's testimony, that the proximity of death could actually enhance lives lived under terror and on the edge of the impossible; and so is the statement by the Belsen survivor printed as an epigraph to this book. Here the Israeli poet Dan Pagis must have the final word, in an uncompleted sentence.

During the plague I came into my own.
It was a time of smoke-pots in the house
Against infection. The blind head of bone
Grinned its abuse

Like a good democrat at everyone.
Runes were recited daily, charms were applied.
That was the time I came into my own.
Half Europe died.

The symptoms are a fever and dark spots
First on the hands, then on the face and neck,
But even before the body, the mind rots.
You can be sick

Only a day with it before you're dead.
But the most curious part of it is the dance.
The victim goes, in short, out of his head.
A sort of trance

Glazes the eyes, and then the muscles take
His will away from him, the legs begin
Their funeral jig, the arms and belly shake
Like souls in sin.

Some, caught in these convulsions, have been known
To fall from windows, fracturing the spine.
Others have drowned in streams. The smooth head-stone,
The box of pine,

Are not for the likes of these. Moreover, flame
Is powerless against contagion.
That was the black winter when I came
Into my own.

ANTHONY HECHT (b. 1923), 'Tarantula, or The Dance of Death'

He that durst in the dead hour of gloomy midnight have been so valiant as to have walked through the still and melancholy streets—what think you should have been his music? Surely the loud groans of raving sick men, the struggling pangs of souls departing; in every house grief striking up an alarum—servants crying out for masters, wives for husbands, parents for children, children for their mothers. Here, he should have met some franticly running to knock up sextons; there, others fearfully sweating with coffins to steal forth dead bodies lest the fatal handwriting of Death should seal up their doors. And to make this dismal concert more full, round about him bells heavily tolling in one place and ringing out in another. The dreadfulness of such an hour is unutterable . . .

Let us look forth and try what consolation rises with the sun. Not any, not any. For before the jewel of the morning be fully set in silver, a hundred hungry graves stand gaping and every one of them, as at a breakfast, hath swallowed down ten or eleven lifeless carcasses. Before dinner in the same gulf are twice so many more devoured. And before the sun takes his rest those numbers are doubled. Threescore, that not many hours before had every one several lodgings very delicately furnished, are now thrust all together into one close room, a little, little, noisome room not fully ten foot square . . .

Imagine then that all this while Death, like a Spanish leaguer—or rather, like stalking Tamburlaine—hath pitched his tents, being nothing but a heap of winding-sheets tacked together, in the sinfully polluted suburbs. The plague is muster-master and marshal of the field; burning fevers, boils, blains and carbuncles the leaders, lieutenants, sergeants and corporals; the main army consisting like Dunkirk of a mingle-mangle, viz. dumpish mourners, merry sextons, hungry coffin-sellers, scrubbing bearers and nasty grave-makers (but indeed they are the pioneers of the camp, that are employed only like moles in casting up of earth and digging of trenches); fear and trembling, the two catch-poles of Death, arrest everyone.

THOMAS DEKKER (?1572–?1632), *The Wonderful Year*, 1603

At last I fell into some talk, at a distance, with this poor man; first I asked him how people did thereabouts. 'Alas, sir!' says he, 'almost desolate; all dead or sick. Here are very few families in this part, or in that village' (pointing at Poplar), 'where half of them are not dead already, and the rest sick.' Then he pointing to one house, 'There they are all dead,' said he, 'and the house stands open; nobody dares go into it. A poor thief,' says he, 'ventured in to steal something, but he paid dear for his theft, for he was carried to the churchyard too last night.' Then he pointed to several other houses. 'There,' says he, 'they are all dead, the man and his wife, and five children. There,' says he, 'they are shut up; you see a watchman at the door'; and so of other houses. 'Why,' says I, 'what do you here all alone?' 'Why,' says he, 'I am a poor desolate man; it has pleased God I am not yet visited, though my family is, and one of my children dead.' 'How do you mean, then,' said I, 'that you are not visited?' 'Why,' says he, 'that's my house' (pointing to a very little, low-boarded house), 'and there my poor wife and two children live,' said he, 'if they may be said to live, for my wife and one of the children are visited, but I do not come at them.' And with that word I saw the tears run very plentifully down his face; and so they did down mine too, I assure you.

'But,' said I, 'why do you not come at them? How can you abandon your own flesh and blood?' 'Oh sir,' says he, 'the Lord forbid! I do not abandon them; I work for them as much as I am able; and, blessed be the Lord, I keep them from want'; and with that I observed he lifted up his eyes to heaven, with a countenance that presently told me I had happened on a man that was no hypocrite, but a serious, religious, good man, and his ejaculation was an expression of thankfulness that, in such a condition as he was in, he should be able to say his family did not want. 'Well,' says I, 'honest man, that is a great mercy as things go now with the poor. But how do you live, then, and how are you kept from the dreadful calamity that is now upon us all?' 'Why, sir,' says he, 'I am a waterman, and there's my boat,' says he, 'and the boat serves me for a house. I work in it in the day, and I sleep in it in the night; and what I get I lay down upon that stone,' says he, showing me a broad stone on the other side of the street, a good way from his house; 'and then,' says he, 'I halloo, and call to them till I make them hear; and they come and fetch it.'

DANIEL DEFOE (?1660–1731), *A Journal of the Plague Year, 1665*

August 31st. Thus this month ends with great sadness upon the public, through the greatness of the plague everywhere through the kingdom almost. Every day sadder and sadder news of its increase. In the City died this week 7496, and of them 6102 of the plague. But it is feared that the true number of the dead this week is near 10,000; partly from the poor that cannot be taken notice of, through the greatness of the number, and partly from the Quakers and others that will not have any bell ring for them . . .

September 3 (Lord's Day). Up; and put on my coloured silk suit very fine, and my new periwig, bought a good while since, but durst not wear, because the plague was in Westminster when I bought it; and it is a wonder what will be the fashion after the plague is done, as to periwigs, for nobody will dare to buy any hair, for fear of the infection, that it had been cut off the heads of people dead of the plague.

SAMUEL PEPYS, *Diary*, 1665

One evening Cottard and Tarrou went to the Municipal Theatre and Opera House, where Gluck's *Orpheus* was being given . . . A touring operatic company had come to Oran in the spring for a short series of performances of this opera. Marooned there by the outbreak of plague and finding themselves in difficulties, the company had come to an agreement with the management of the Opera House, under which they were to give one performance a week until further notice . . .

Throughout the first act Orpheus lamented suavely his lost Eurydice, with women in Grecian tunics singing melodious comments on his plight, and love was hymned in alternating arias. The audience showed their appreciation in discreet applause. Only a few people noticed that in his song of the Second Act Orpheus introduced some tremolos not in the score, and voiced an almost exaggerated emotion when begging the Lord of the Shades to be moved by his tears. Some rather jerky movements he indulged in gave our connoisseurs of stagecraft an impression of clever, if slightly overdone, effects, intended to bring out the emotion of the words he sang.

Not until the big duet between Orpheus and Eurydice in the Third Act—at the precise moment when Eurydice was being torn from her lover—did a flutter of surprise run through the house. And, as though the singer had been waiting for this cue, or, more likely, because the faint sounds that came to him from stalls and pit confirmed what he was feeling, he chose this moment to stagger grotesquely to the footlights, his arms and legs splayed out under his antique robe, and fall down in the middle of the property sheepfold, always anachronistic, but now in the eyes of the spectators, significantly, appallingly so. For at the same moment the orchestra stopped playing, the audience rose and began to leave the auditorium, slowly and silently at first, like worshippers leaving church when the service ends, or a death-chamber after a farewell visit to the dead, women lifting their skirts and moving with bowed heads, men steering the ladies by the elbow to prevent their brushing against the tip-up seats at the ends of the rows. But gradually their movements quickened, whispers rose to exclamations, and finally the crowd stampeded towards the exits, wedged together in the bottle-necks, and poured out into the street in a confused mass, with shrill cries of dismay.

CAMUS, *The Plague*, tr. Stuart Gilbert

They fought south of the ramparts,
They died north of the wall.
They died in the moors and were not buried.
Their flesh was the food of crows.
'Tell the crows we are not afraid;
We have died in the moors and cannot be buried.
Crows, how can our bodies escape you?'
The waters flowed deep
And the rushes in the pool were dark.
The riders fought and were slain;
Their horses wander neighing.
By the bridge there was a house.
Was it south, was it north?
The harvest was never gathered.
How can we give you your offerings?
You served your Prince faithfully,
Though all in vain.
I think of you, faithful soldiers;
Your service shall not be forgotten.
For in the morning you went out to battle
And at night you did not return.

Chinese (?1st century), 'Fighting South of the Ramparts', tr. Waley

Take this message to the Spartans, passer-by:
Obedient to their orders, here we lie.

SIMONIDES (556–467 BC) (on the Spartan dead at Thermopylae)

If any question why we died,
Tell them, because our fathers lied.

KIPLING, 'Common Form'

Tamburlaine. Virgins, in vain you labour to prevent
That which mine honour swears shall be perform'd.
Behold my sword; what see you at the point?
First Virgin. Nothing but fear and fatal steel, my lord.
Tamb. Your fearful minds are thick and misty, then,
For there sits death; there sits imperious Death,
Keeping his circuit by the slicing edge.
But I am pleas'd you shall not see him there;
He now is seated on my horsemen's spears,
And on their points his fleshless body feeds.—
Techelles, straight go charge a few of them
To charge these dames, and show my servant Death,
Sitting in scarlet on their armèd spears.

MARLOWE, *Tamburlaine the Great*

The body of a dead enemy always smells good.

CHARLES IX OF FRANCE (1550–74)

When a man hath no freedom to fight for at home,
Let him combat for that of his neighbours;
Let him think of the glories of Greece and of Rome,
And get knocked on his head for his labours.

To do good to mankind is the chivalrous plan,
And is always as nobly requited;
Then battle for freedom wherever you can,
And, if not shot or hanged, you'll get knighted.

LORD BYRON (1788–1824), 'Stanzas'

Prince Andrew's regiment was among the reserves which till after one o'clock were stationed inactive behind Semënovsk under heavy artillery fire. Towards two o'clock the regiment, having already lost more than two hundred men, was moved forward into a trampled oatfield in the gap between Semënovsk and the Knoll Battery where thousands of men perished that day and on which an intense, concentrated fire from several hundred enemy guns was directed between one and two o'clock.

Without moving from that spot or firing a single shot the regiment here lost another third of its men. From in front and especially from the right, in the unlifting smoke the guns boomed, and out of the mysterious domain of smoke that overlay the whole space in front, quick hissing cannon-balls and slow whistling shells flew unceasingly. At times, as if to allow them a respite, a quarter of an hour passed during which the cannon-balls and shells all flew overhead, but sometimes several men were torn from the regiment in a minute, and the slain were continually being dragged away and the wounded carried off . . .

'Look out!' came a frightened cry from a soldier and, like a bird whirring in rapid flight and alighting on the ground, a shell dropped with little noise within two steps of Prince Andrew and close to the battalion commander's horse. The horse first, regardless of whether it was right or wrong to show fear, snorted, reared almost throwing the major, and galloped aside. The horse's terror infected the men.

'Lie down!' cried the adjutant, throwing himself flat on the ground.

Prince Andrew hesitated. The smoking shell spun like a top between him and the prostrate adjutant near a wormwood plant between the field and the meadow.

'Can this be death?' thought Prince Andrew, looking with a quite new, envious glance at the grass, the wormwood, and the streamlet of smoke that curled up from the rotating black ball. 'I cannot, I do not wish to die. I love life—I love this grass, this earth, this air . . .' He thought this, and at the same time remembered that people were looking at him.

'It's shameful, sir!' he said to the adjutant. 'What . . .'

He did not finish speaking. At one and the same moment came the sound of an explosion, a whistle of splinters as from a breaking window frame, a suffocating smell of powder, and Prince Andrew started to one side, raising his arm, and fell on his chest. Several officers ran up to him. From the right side of his abdomen blood was welling out making a large stain on the grass.

TOLSTOY, *War and Peace*, tr. Louise and Aylmer Maude

When the furious struggle of the present war has been decided, each one of the victorious fighters will return home joyfully to his wife and children, unchecked and undisturbed by thoughts of the enemies he has killed whether at close quarters or at long range. It is worthy of note that the primitive races which still survive in the world, and are undoubtedly closer than we are to primeval man, act differently in this respect, or did until they came under the influence of our civilization. Savages . . . are far from being remorseless murderers; when they return victorious from the war-path they may not set foot in their villages or touch their wives till they have atoned for the murders they committed in war by penances which are often long and tedious. It is easy, of course, to attribute this to their superstition: the savage still goes in fear of the avenging spirits of the slain. But the spirits of his slain enemy are nothing but the expression of his bad conscience about his blood-guilt; behind this superstition there lies concealed a vein of ethical sensitiveness which has been lost by us civilized men.

FREUD, 'Thoughts for the Times on War and Death', 1915

An old reservist looked at the raw recruit and said: '. . . They've pulled the wool over our eyes. Once a deputy from the Clerical Party came to our village and spoke to us about God's peace, which spans the earth, and how the Lord did not want war and wanted us all to live in peace and get on together like brothers. And look at him now, the bloody fool! Now that war has broken out they pray in all the churches for the success of our arms, and they talk about God like a chief of the general staff who guides and directs the war. From this military hospital I've seen many funerals go out and cartfuls of hacked-off arms and legs carried away.'

'And the soldiers are buried naked,' said another soldier, 'and into the uniform they put another live man. And so it goes on for ever and ever.'

'Until we've won,' observed Švejk.

'And that bloody half-wit wants to win something,' a corporal chimed in from the corner. 'To the front with you, to the trenches! You should be driven for all you're worth on to bayonets over barbed wire, mines and mortars. Anyone can lie about behind the lines, but no one wants to fall in action.'

'I think that it's splendid to get oneself run through with a bayonet,' said Švejk, 'and also that it's not bad to get a bullet in the stomach. It's even grander when you're torn to pieces by a shell and you see that your legs and belly are somehow remote from you. It's very funny and you die before anyone can explain it to you.'

Wiping from his chin the drops of sauce and fat which fell from the bread he continued: 'I really can't imagine what you'd have done without me if they had held me somewhere else and the war had continued a few years longer.'

Vaněk asked with interest: 'How long do you think the war will go on, Švejk?'

'Fifteen years,' answered Švejk. 'That's obvious because once there was a thirty years' war and now we're twice as clever as they were before, so it follows that thirty divided by two is fifteen.'

JAROSLAV HAŠEK (1883–1923), *The Good Soldier Švejk*, tr. Cecil Parrott

None saw their spirits' shadow shake the grass,
Or stood aside for the half used life to pass
Out of those doomed nostrils and the doomed mouth,
When the swift iron burning bee
Drained the wild honey of their youth.

ISAAC ROSENBERG (1890–1918), from 'Dead Man's Dump'

Who died on the wires, and hung there, one of two—
Who for his hours of life had chattered through
Infinite lovely chatter of Bucks accent:
Yet faced unbroken wires; stepped over, and went
A noble fool, faithful to his stripes—and ended . . .

GURNEY, from 'The Silent One'

We were among the debris of the intense bombardment of ten days before, for we were passing along and across the Hindenburg Outpost Trench, with its belt of wire (fifty yards deep in places); here and there these rusty jungles had been flattened by tanks. The Outpost Trench was about 200 yards from the Main Trench, which was now our front line. It had been solidly made, ten feet deep, with timbered fire-steps, splayed sides, and timbered steps at intervals to front and rear and to machine-gun emplacements. Now it was wrecked as though by earthquake and eruption. Concrete strong-posts were smashed and tilted sideways; everywhere the chalky soil was pocked and pitted with huge shell-holes; and wherever we looked the mangled effigies of the dead were our *memento mori*. Shell-twisted and dismembered, the Germans maintained the violent attitudes in which they had died. The British had mostly been killed by bullets or bombs, so they looked more resigned. But I can remember a pair of hands (nationality unknown) which protruded from the soaked ashen soil like the roots of a tree turned upside down; one hand seemed to be pointing at the sky with an accusing gesture. Each time I passed that place the protest of those fingers became more expressive of an appeal to God in defiance of those who made the War. Who made the War? I laughed hysterically as the thought passed through my mud-stained mind. But I only laughed mentally, for my box of Stokes gun ammunition left me no breath to spare for an angry guffaw. And the dead were the dead; this was no time to be pitying them or asking silly questions about their outraged lives. Such sights must be taken for granted, I thought, as I gasped and slithered and stumbled with my disconsolate crew. Floating on the surface of the flooded trench was the mask of a human face which had detached itself from the skull.

SIEGFRIED SASSOON (1886–1967), *Memoirs of an Infantry Officer*

Hooded in angry mist, the sun goes down:
Steel-grey the clouds roll out across the sea;
Is this a Kingdom? Then give Death the crown,
For here no emperor hath won, save He.

HERBERT ASQUITH (1881–1947), 'Nightfall'

I know that I shall meet my fate
Somewhere among the clouds above;
Those that I fight I do not hate,
Those that I guard I do not love;
My country is Kiltartan Cross,
My countrymen Kiltartan's poor,
No likely end could bring them loss
Or leave them happier than before.
Nor law, nor duty bade me fight,
Nor public men, nor cheering crowds,
A lonely impulse of delight
Drove to this tumult in the clouds;
I balanced all, brought all to mind,
The years to come seemed waste of breath,
A waste of breath the years behind
In balance with this life, this death.

YEATS, 'An Irish Airman foresees his Death'

A whole night through
thrown down beside
a butchered comrade
with his clenched teeth
turned to the full moon
and the clutching
of his hands
thrust
into my silence
I have written
letters full of love

Never have I
clung
so fast to life

GIUSEPPE UNGARETTI (1888–1970), 'Watch, Cima Quattro, 23 December 1915', tr. Patrick Creagh

Out there, we've walked quite friendly up to Death;
Sat down and eaten with him, cool and bland,—
Pardoned his spilling mess-tins in our hand.
We've sniffed the green thick odour of his breath,—
Our eyes wept, but our courage didn't writhe.
He's spat at us with bullets and he's coughed
Shrapnel. We chorused when he sang aloft;
We whistled while he shaved us with his scythe.

Oh, Death was never enemy of ours!
We laughed at him, we leagued with him, old chum.
No soldier's paid to kick against his powers.
We laughed, knowing that better men would come,
And greater wars; when each proud fighter brags
He wars on Death—for lives; not men—for flags.

OWEN, 'The Next War'

From my mother's sleep I fell into the State,
And I hunched in its belly till my wet fur froze.
Six miles from earth, loosed from its dream of life,
I woke to black flak and the nightmare fighters.
When I died they washed me out of the turret with a hose.

RANDALL JARRELL (1914–65), 'The Death of the Ball Turret Gunner'

There was one maintenance man who made a point of meticulously scouring and polishing the cockpit of each kamikaze plane he tended. It was his theory that the cockpit was the pilot's coffin and as such it should be spotless. One recipient of this service was so pleasantly surprised that he summoned and thanked his benefactor, saying that the neatness of the plane meant a great deal to him. The maintenance man's eyes dimmed with tears, and, unable to speak, he ran along with one hand on the wing tip of the plane as it taxied for its final take-off.

NAKAJIMA TADASHI, *The Divine Wind: Japan's Kamikaze Force in World War II*, tr. Roger Pineau, 1958

In Pilsen,
Twenty-six Station Road,
she climbed to the Third Floor
up stairs which were all that was left
of the whole house,
she opened her door
full on to the sky,
stood gaping over the edge.

For this was the place
the world ended.

Then
she locked up carefully
lest someone steal
Sirius
or Aldebaran
from her kitchen,
went back downstairs
and settled herself
to wait
for the house to rise again
and for her husband to rise from the ashes
and for her children's hands and feet to be stuck
back in place.

In the morning they found her
still as stone,
sparrows pecking her hands.

MIROSLAV HOLUB (b. 1923), 'Five Minutes After the Air Raid', tr. Ian Milner and George Theiner

Here are two pictures from my father's head—
I have kept them like secrets until now:
First, the Ulster Division at the Somme
Going over the top with 'Fuck the Pope!'
'No Surrender': a boy about to die,
Screaming 'Give 'em one for the Shankhill!'
'Wilder than Gurkhas' were my father's words
Of admiration and bewilderment.
Next comes the London-Scottish padre
Resettling kilts with his swagger-stick,
With a stylish backhand and a prayer.
Over a landscape of dead buttocks
My father followed him for fifty years.
At last, a belated casualty,
He said—lead traces flaring till they hurt—
'I am dying for King and Country, slowly.'
I touched his hand, his thin head I touched.

Now, with military honours of a kind,
With his badges, his medals like rainbows,
His spinning compass, I bury beside him
Three teenage soldiers, bellies full of
Bullets and Irish beer, their flies undone.
A packet of Woodbines I throw in,
A lucifer, the Sacred Heart of Jesus
Paralysed as heavy guns put out
The night-light in a nursery for ever;
Also a bus-conductor's uniform—
He collapsed beside his carpet-slippers
Without a murmur, shot through the head
By a shivering boy who wandered in
Before they could turn the television down
Or tidy away the supper dishes.
To the children, to a bewildered wife,
I think 'Sorry Missus' was what he said.

MICHAEL LONGLEY (b. 1939), 'Wounds'

The canal was full of bodies: I am reminded now of an Irish stew containing too much meat. The bodies overlapped: one head, seal-grey, and anonymous as a convict with a shaven scalp, stuck up out of the water like a buoy. There was no blood: I suppose it had flowed away a long time ago. I have no idea how many there were: they must have been caught in a cross-fire, trying to get back, and I suppose every man of us along the bank was thinking, 'Two can play at that game.' I too took my eyes away; we didn't want to be reminded of how little we counted, how quickly, simply and anonymously death came. Even though my reason wanted the state of death, I was afraid like a virgin of the act. I would have liked death to come with due warning, so that I could prepare myself. For what? I didn't know, nor how, except by taking a look around at the little I would be leaving.

GRAHAM GREENE (b. 1904), *The Quiet American*

The mother does knitting
The son goes to the war
She finds this perfectly natural the mother
And the father what does he do the father
He does business
His wife does knitting
His son goes to the war
Himself he does business
He finds this perfectly natural the father
And the son the son
What does the son find the son
He finds nothing he finds nothing at all
His mother does knitting his father does business he goes to the war
When he has finished the war
He will go into business along with his father
The war goes on the mother goes on she does knitting
The father goes on he does business
The son is killed he goes on no further
The father and the mother go to the cemetery
They find this natural the father and the mother
Life goes on with knitting with war with business
With business and war and knitting and war
And business and business and business
Life goes on with the cemetery.

JACQUES PRÉVERT (1900–77), 'Familial'

Ring-a-ring o' neutrons
A pocket full of positrons,
A fission! A fission!
We all fall down.

PAUL DEHN (1912–76)

'A planet doesn't explode of itself,' said drily
The Martian astronomer, gazing off into the air—
'That they were able to do it is proof that highly
Intelligent beings must have been living there.'

JOHN HALL WHEELOCK (1886–1978), 'Earth'

Days of misfortune pass and are gone,
Like the days of winter, they come and they go;
The sorrows of men do not last very long,
Like the buyers in shops, they come and they go.

Persecution and blood lash the people to tears,
The caravans, they come and they go;
And men spring up in the garden of earth,
Whether henbane or balsam, they come and they go . . .

Armenian folk-song, in Franz Werfel, *The Forty Days of Musa Dagh*, 1933, tr. Geoffrey Dunlop

I'm told you raised your hand against yourself
Anticipating the butcher.
After eight years in exile, observing the rise of the enemy
Then at last, brought up against an impassable frontier
You passed, they say, a passable one.

Empires collapse. Gang leaders
Are strutting about like statesmen. The peoples
Can no longer be seen under all those armaments.

So the future lies in darkness and the forces of right
Are weak. All this was plain to you
When you destroyed a torturable body.

BERTOLT BRECHT (1898–1956), 'On the Suicide of the Refugee W. B.', tr. John Willett

Many people were arrested; most were very simpleminded—not clever enough to plead innocent—but the executioners themselves seemed frightened by all the killing . . . Every day, one or two hundred innocent farmers were arrested.

The officials would not let them go, nor did they want to kill all of them. The dilemma was soon solved by arranging a procedure for selection, the responsibility for which was assigned to the Heavenly King worshipped by the local people. The soldiers led the prisoners to the temple, to the main chamber, where each was made to cast a pair of bamboo rods: one face up and one face down was 'regular', and the prisoner was released; both faces up meant one was *yang* fodder, and also meant release; both faces down meant *yin* fodder, and that the prisoner was condemned to die. Life or death depended on the cast. Those who were to die went to the left; those who were to live went to the right. A man who had been given two chances out of three to live remained silent when it was determined he had to die; he lowered his head and walked to the left.

SHEN TS'UNG-WEN, on an uprising in Fenghuang in 1911, tr. Nieh Hua-ling

And another thing: death. I was told about a camp where transports of new prisoners arrived each day, dozens of people at a time. But the camp had only a certain quantity of daily food rations—I cannot recall how much, maybe enough for two, maybe three thousand—and Herr Kommandant disliked to see the prisoners starve. Each man, he felt, must receive his allotted portion. And always the camp had a few dozen men too many. So every evening a ballot, using cards or matches, was held in every block, and the following morning the losers did not go to work. At noon they were led out behind the barbed-wire fence and shot.

TADEUSZ BOROWSKI, 'Auschwitz, Our Home: A Letter'

There is a country Lost,
a moon grows in its reeds,
where all that died of frost
as we did, glows and sees.

It sees, for it has eyes,
each eye an earth, and bright.
The night, the night, the lyes.
This eye-child's gift is sight.

It sees, it sees, we see,
I see you, you see me.
Before this hour has ended
Ice will rise from the dead.

PAUL CELAN (1920–70), 'Ice, Eden', tr. Michael Hamburger

Be happy, you who live in fine apartments, in ugly houses or in hovels. Be happy, you who have your loved ones, and you also who sit alone and dream and can weep. Be happy, you who torture yourselves over metaphysical problems, and you who suffer because of money worries. Be happy, you the sick who are being cared for, and you who care for them, and be happy, oh how happy, you who die a death as normal as life, in hospital beds or in your homes. Be happy, all of you: millions of people envy you.

MICHELINE MAUREL, a survivor of the Nazi camp, *Ravensbrück*, tr. Margaret S. Summers

Our way of life kept us firmly rooted to the ground, and was not conducive to the search for transcendental truths. Whenever I talked of suicide, M. used to say: 'Why hurry? The end is the same everywhere, and here they even hasten it for you.' Death was so much more real, so much simpler than life, that we all involuntarily tried to prolong our earthly existence, even if only for a brief moment—just in case the next day brought some relief! In war, in the camps and during periods of terror, people think much less about death (let alone suicide) than when they are living normal lives. Whenever at some point on earth mortal terror and the pressure of utterly insoluble problems are present in a particularly intense form, general questions about the nature of being recede into the background . . . In a strange way, despite the horror of it, this also gave a certain richness to our lives.

NADEZHDA MANDELSTAM, *Hope Against Hope*, tr. Max Hayward, 1970

here in this carload
i am eve
with abel my son
if you see my other son
cain son of man
tell him i

DAN PAGIS (b. 1930), 'Written in Pencil in the Sealed Railway-Car', tr. Stephen Mitchell

Love and Death

Who lets so fair a house fall to decay,
Which husbandry in honour might uphold,
Against the stormy gusts of winter's day
And barren rage of death's eternal cold?
O none but unthrifts. Dear my love you know,
You had a father, let your son say so.

SHAKESPEARE's friend once had a father; now he must go about having a son, and in his turn ward off the 'barren rage' of death. Although Arthur Koestler considers the dichotomy between Eros and Thanatos too exclusive and simple-minded—'the great duet in Freud's metapsychology does not constitute the whole opera'—many before and since Freud have seen the two polar opposites as 'primal instincts' and ourselves as the outcome of the war or the stalemate or the compromise between them.

'Yes, they are carnal, both of them, love and death'—*l'amour et la mort*—'and there lie their terror and their great magic!' Thus Hans Castorp in the excitement of carnival at the sanatorium on the Magic Mountain. Others have conflated the two into one: here, Unamuno and, in a more spiritual sense, Gottfried von Strassburg in celebrating the influential story of Tristan and Isolde. In another of his sonnets Shakespeare invokes love as a preservative against the death wish: tired of the world's inequities, 'captive good attending captain ill', he would gladly be gone, 'save that, to die, I leave my love alone'.

That men and women have ceased to live when they lose the one they have long loved is beyond doubt, but a question of popular debate is whether or not people really kill themselves for love. Rosalind's famous diatribe—all those old tales are lies, she contends—is not to be taken too seriously, since she is bracing up the love-sick Orlando, and herself as well. The passage from the seventeenth-century Samuel Butler is rather more interesting than it seems at first sight: he does not dispute the existence of love-suicide but claims that it is for ill-natured women, not the tender-hearted and more deserving, that men kill themselves. Probably Shakespeare would not agree wholeheartedly with either Rosalind or Butler; and nor, I imagine, do most of us, whatever our views on the propriety of suicide.

The suicide pact between Kleist and Henriette Vogel was of its own kind, deriving chiefly from his philosophical anguish and her fatal sickness. But I remember a pact discharged in a hotel in Johore Bahru, between an Indian boy and a Chinese girl, both of them well educated and intelligent, both forbidden by their families to marry outside the race, a pact only marred in its symmetry by the accident that the girl's overdose proved fatal and the boy's did not.

Mystical love in its relation to death is touched on in its most famous exemplar, St Teresa of Avila, and in the Corpus Christi Carol. At the other end of the scale, apart from a few velleities and fancies, I have avoided literal 'love of the dead' except for a case of fortunate (and false) necrophilia and an instance which, being classical, is harmless and even in the sequel gentlemanly. Incidentally, where Graves follows the legend, Kleist (as cited in the passage on the vampire's kiss) reverses it, so that Penthesilea kills Achilles under a misapprehension.

In his judgement between Love and Death the youthful Tennyson's award of the palm to the former is quaintly theatrical and more pietistic than persuasive. Yet the preponderance of the evidence is in his favour, and accords with another of Shakespeare's sonnets:

> Love's not Time's fool, though rosy lips and cheeks
> Within his bending sickle's compass come;
> Love alters not with his brief hours and weeks,
> But bears it out even to the edge of doom . . .

By definition, any love that fails to live up to this victory over death—or at least this draw with it—is not love.

> What time the mighty moon was gathering light
> Love paced the thymy plots of Paradise,
> And all about him roll'd his lustrous eyes;
> When, turning round a cassia, full in view,
> Death, walking all alone, beneath a yew,
> And talking to himself, first met his sight:
> 'You must begone,' said Death, 'these walks are mine.'
> Love wept and spread his sheeny vans for flight;
> Yet ere he parted said, 'This hour is thine:
> Thou art the shadow of life, and as the tree
> Stands in the sun and shadows all beneath,
> So in the light of great eternity
> Life eminent creates the shade of death;
> The shadow passeth when the tree shall fall,
> But I shall reign for ever over all.'

TENNYSON, 'Love and Death'

To live is to give oneself, perpetuate oneself, and to perpetuate oneself, to give oneself, is to die. Perhaps the supreme delight of procreation is nothing other than a foretasting or savouring of death, the spilling of one's own vital essence. We unite with another, but it is to divide ourselves: that most intimate embrace is naught but a most intimate uprooting. In essence, the delight of sexual love, the genetic spasm, is a sensation of resurrection, of resuscitation in another, for only in others can we resuscitate and perpetuate ourselves.

UNAMUNO, *The Tragic Sense of Life*

For us their life, their death, are bread,
They live as living, they live as dead,
Thus still they live and yet are dead,
And for the living their death is bread.

GOTTFRIED VON STRASSBURG, *Tristan und Isolt*, *c.*1210

Above the horizon the bank of cloud divided, silver in grey. Far out on the waves, a ship was speeding towards the southwest. Long, Marc gazed at the swelling sail.

They have wasted their lives, he thought; they have burned themselves out; yet they lived not in vain. They were flame, that the earth grow not cold by their burning . . .

I sought to teach men to live, he thought with bitterness, but what they seek is to transcend their death.

HANNAH CLOSS, *Tristan*, 1940

In me is death, in you my life.
You mark, concede, divide up time;
Or long or short my life, as you wish.

Happy am I in your kindness.
Thanks to you the soul is blest,
And contemplates God where no time is.

MICHELANGELO (1475–1564)

Lully, lullay, lully, lullay,
The falcon hath borne my make away.

He bore him up, he bore him down;
He bore him into an orchard brown.

In that orchard there was a hall,
That was hanged with purple and pall.

And in that hall there was a bed,
It was hanged with gold so red.

And in that bed there lieth a knight,
His woundës bleeding day and night.

By that bed's side there kneeleth a may,
And she weepeth both night and day.

And by that bed's side there standeth a stone,
Corpus Christi written thereon.

Corpus Christi Carol, from a 16th-century ms.

THOU art Love's victim; and must die
A death more mystical and high.
Into love's arms thou shalt let fall
A still-surviving funeral.
His is the DART must make the DEATH
Whose stroke shall taste thy hallow'd breath;
A Dart thrice dipp'd in that rich flame
Which writes thy spouse's radiant Name
Upon the roof of Heav'n . . .
So rare,
So spiritual, pure, and fair
Must be th'immortal instrument
Upon whose choice point shall be sent
A life so lov'd. And that there be
Fit executioners for thee,
The fair'st and first-born sons of fire
Blest SERAPHIM, shall leave their quire
And turn love's soldiers, upon THEE
To exercise their archery.

O how oft shalt thou complain
Of a sweet and subtle PAIN;
Of intolerable JOYS;
Of a DEATH, in which who dies
Loves his death, and dies again;
And would for ever so be slain.
And lives, and dies; and knows not why
To live, but that he thus may never leave to DIE.
How kindly will thy gentle HEART
Kiss the sweetly-killing DART!
And close in his embraces keep
Those delicious Wounds, that weep
Balsam to heal themselves with. Thus
When These thy DEATHS, so numerous,
Shall all at last die into one,
And melt thy Soul's sweet mansion;
Like a soft lump of incense, hasted
By too hot a fire, and wasted
Into perfuming clouds, so fast
Shalt thou exhale to Heav'n at last
In a resolving SIGH, and then
O what? Ask not the Tongues of men.
Angels cannot tell, suffice,
Thy self shall feel thine own full joys
And hold them fast for ever.

RICHARD CRASHAW (1612/13–49), from 'Hymn to Saint Teresa'

Sleep on my Love in thy cold bed
Never to be disquieted!
My last good night! Thou wilt not wake
Till I thy fate shall overtake:
Till age, or grief, or sickness, must
Marry my body to that dust
It so much loves; and fill the room
My heart keeps empty in thy Tomb.
Stay for me there; I will not fail
To meet thee in that hollow Vale.
And think not much of my delay;
I am already on the way,
And follow thee with all the speed
Desire can make, or sorrows breed . . .

'Tis true, with shame and grief I yield,
Thou like the Van first took'st the field,
And gotten hast the victory
In thus adventuring to die
Before me, whose more years might crave
A just precedence in the grave.
But hark! My pulse like a soft drum
Beats my approach, tells thee I come;
And slow howe'er my marches be,
I shall at last sit down by thee.

HENRY KING (1592–1669), from 'The Exequy

To these, whom Death again did wed,
This grave's their second marriage-bed.
For though the hand of fate could force
'Twixt soul and body a divorce,
It could not sunder man and wife,
'Cause they both livèd but one life.
Peace, good Reader. Do not weep.
Peace, the lovers are asleep.
They, sweet turtles, folded lie
In the last knot love could tie;
And though they lie as they were dead,
Their pillow stone, their sheets of lead,
(Pillow hard, and sheets not warm)
Love made the bed; they'll take no harm.
Let them sleep: let them sleep on,
Till this stormy night be gone,
Till the eternal morrow dawn;
Then the curtains will be drawn
And they wake into a light
Whose day shall never die in night.

CRASHAW, 'An Epitaph upon a Young Married Couple, Dead and Buried Together'

Then thus the sire of gods, with look serene,
'Speak thy desire, thou only just of men;
And thou, O woman, only worthy found
To be with such a man in marriage bound.'

A while they whisper; then, to Jove address'd,
Philemon thus prefers their joint request:
'We crave to serve before your sacred shrine,
And offer at your altars rites divine:
And since not any action of our life
Has been polluted with domestic strife,
We beg one hour of death; that neither she
With widow's tears may live to bury me,
Nor weeping I, with wither'd arms may bear
My breathless Baucis to the sepulcher.'
The godheads sign their suit. They run their race
In the same tenor all th' appointed space;
Then, when their hour was come, while they relate
These past adventures at the temple-gate,
Old Baucis is by old Philemon seen
Sprouting with sudden leaves of spritely green:
Old Baucis look'd where old Philemon stood,
And saw his lengthen'd arms a sprouting wood:
New roots their fasten'd feet begin to bind,
Their bodies stiffen in a rising rind:
Then e'er the bark above their shoulders grew,
They give and take at once their last adieu;
At once, farewell, O faithful spouse, they said;
At once th' incroaching rinds their closing lips invade.
Ev'n yet, an ancient Tyanaean shows
A spreading oak, that near a linden grows:
The neighbourhood confirm the prodigie,
Grave men, not vain of tongue, or like to lie.

OVID (43 BC–?AD 17), *Metamorphoses*, tr. Dryden

But if you survived melancholia and rotting lungs it was possible to live long in this valley. Joseph and Hannah Brown, for instance, appeared to be indestructible. For as long as I could remember they had lived together in the same house by the common . . . They had raised a large family and sent them into the world, and had continued to live on alone, with nothing left of their noisy brood save some dog-eared letters and photographs.

It seemed that the old Browns belonged for ever, and that the miracle of their survival was made commonplace by the durability of their love—if one should call it love, such a balance. Then suddenly, within the space of two days, feebleness took them both. It was as though two machines, wound up and synchronized, had run down at exactly the same time. Their interdependence was so legendary we didn't notice their plight at first. But after a week, not having been seen about, some neighbours thought it best to call. They found old Hannah on the kitchen floor feeding her man with a spoon. He was lying in a corner half-covered with matting, and they were both too weak to stand. She had chopped up a plate of peelings, she said, as she hadn't been able to manage the fire. But they were all right really, just a touch of the damp; they'd do, and it didn't matter.

Well, the Authorities were told; the Visiting Spinsters got busy; and it was decided they would have to be moved. They were too frail to help each other now, and their children were too scattered, too busy. There was but one thing to be done; it was for the best; they would have to be moved to the Workhouse.

The old couple were shocked and terrified, and lay clutching each other's hands. 'The Workhouse'—always a word of shame, grey shadow falling on the close of life, most feared by the old (even when called The Infirmary); abhorred more than debt, or prison, or beggary, or even the stain of madness.

Hannah and Joseph thanked the Visiting Spinsters but pleaded to be left at home, to be left as they wanted, to cause no trouble, just simply to stay together. The Workhouse could not give them the mercy they needed, but could only divide them in charity. Much better to hide, or die in a ditch, or to starve in one's familiar kitchen, watched by the objects one's life had gathered—the scrubbed empty table, the plates and saucepans, the cold grate, the white stopped clock . . .

'You'll be well looked after,' the Spinsters said, 'and you'll see each other twice a week.' The bright busy voices cajoled with authority and the old couple were not trained to defy them. So that same afternoon, white and speechless, they were taken away to the Workhouse. Hannah Brown was put to bed in the Women's Wing, and Joseph lay in the Men's. It was the first time, in all their fifty years, that they had ever been separated. They did not see each other again, for in a week they both were dead.

LAURIE LEE (b. 1914), *Cider with Rosie*

He first deceas'd; she for a little tried
To live without him: lik'd it not, and died.

SIR HENRY WOTTON (1568–1639), 'Upon the Death of Sir Albert Morton's Wife'

Not like a suddenly-extinguished light
her spirit left its earthly tenement.
She dwindled like a flamelet, pure and bright,
that lessens in a gradual descent,
keeping its character while waning low,
spending itself, until its force is spent.
Not livid-pale, but whiter than the snow
the hills in windless weather occupying,
only a mortal languor did she show.
She closed her eyes; and in sweet slumber lying,
her spirit tiptoed from its lodging-place.
It's folly to shrink in fear, if this is dying;
for death looked lovely in her lovely face.

PETRARCH, *Triumphs*, tr. Morris Bishop

O my love, my wife!
Death, that hath sucked the honey of thy breath,
Hath had no power yet upon thy beauty.
Thou art not conquered. Beauty's ensign yet
Is crimson in thy lips and in thy cheeks,
And death's pale flag is not advancèd there . . .
Ah, dear Juliet,
Why art thou yet so fair? Shall I believe
That unsubstantial death is amorous,
And that the lean abhorrèd monster keeps
Thee here in dark to be his paramour?
For fear of that I still will stay with thee
And never from this palace of dim night
Depart again . . .
Here's to my love! O true Apothecary!
Thy drugs are quick. Thus with a kiss I die.

SHAKESPEARE, *Romeo and Juliet*

Naples, 1 November [1820]

My dear Brown,

Yesterday we were let out of Quarantine, during which my health suffered more from bad air and the stifled cabin than it had done the whole voyage. The fresh air revived me a little, and I hope I am well enough this morning to write to you a short calm letter;—if that can be called one, in which I am afraid to speak of what I would fainest dwell upon. As I have gone thus far into it, I must go on a little;—

perhaps it may relieve the load of WRETCHEDNESS which presses upon me. The persuasion that I shall see her no more will kill me. I cannot q— [?quit] My dear Brown, I should have had her when I was in health, and I should have remained well. I can bear to die—I cannot bear to leave her. O, God! God! God! Everything I have in my trunks that reminds me of her goes through me like a spear. The silk lining she put in my travelling cap scalds my head. My imagination is horribly vivid about her—I see her—I hear her. There is nothing in the world of sufficient interest to divert me from her a moment. This was the case when I was in England; I cannot recollect, without shuddering, the time that I was a prisoner at Hunt's, and used to keep my eyes fixed on Hampstead all day. Then there was a good hope of seeing her again—Now!—O that I could be buried near where she lives! I am afraid to write to her—to receive a letter from her—to see her handwriting would break my heart—even to hear of her anyhow, to see her name written, would be more than I can bear. My dear Brown, what am I to do? Where can I look for consolation or ease? If I had any chance of recovery, this passion would kill me. Indeed, through the whole of my illness, both at your house and at Kentish-Town, this fever has never ceased wearing me out. When you write to me, which you will do immediately, write to Rome (*poste restante*)—if she is well and happy, put a mark thus +; if— . . . My dear Brown, for my sake, be her advocate for ever. I cannot say a word about Naples; I do not feel at all concerned in the thousand novelties around me. I am afraid to write to her—I should like her to know that I do not forget her. Oh, Brown, I have coals of fire in my breast. It surprises me that the human heart is capable of containing and bearing so much misery. Was I born for this end? God bless her, and her mother, and my sister, and George, and his wife, and you, and all!

Your ever affectionate friend,

JOHN KEATS

Letter to Charles Brown

You've seen a Pair of youthful Lovers die:
And much you care; for, most of you will cry,
'Twas a just Judgement on their Constancy.
For, Heav'n be thanked, we live in such an Age
When no man dies for Love, but on the Stage:
And ev'n those Martyrs are but rare in Plays;
A cursed sign how much true Faith decays.

JOHN DRYDEN (1631–1700), from Epilogue to Nathaniel Lee's *Mithridates*

The poor world is almost six thousand years old, and in all this time there was not any man died in his own person, videlicet, in a love-cause. Troilus had his brains dashed out with a Grecian club, yet he did what he could to die before, and he is one of the patterns of love. Leander, he would have lived many a fair year though Hero had turned nun, if it had not been for a hot mid summer night; for, good youth, he went but forth to wash him in the Hellespont, and being taken with the cramp, was drowned, and the foolish chroniclers of that age found it was Hero of Sestos. But these are all lies: men have died from time to time and worms have eaten them, but not for love.

SHAKESPEARE, *As You Like It*

Romeo and Juliet also embody another popular misconception: that of the suicidal great passion. It seems that those who die for love usually do so by mistake and ill-luck. It is said that the London police can always distinguish, among the corpses fished out of the Thames, between those who have drowned themselves because of unhappy love affairs and those drowned for debt. The fingers of the lovers are almost invariably lacerated by their attempts to save themselves by clinging to the piers of the bridges. In contrast, the debtors apparently go down like slabs of concrete, apparently without struggle and without afterthought.

ALVAREZ, *The Savage God*

Werther had a love for Charlotte
  Such as words could never utter;
Would you know how first he met her?
  She was cutting bread and butter.

Charlotte was a married lady,
  And a moral man was Werther,
And for all the wealth of Indies,
  Would do nothing for to hurt her.

So he sigh'd and pined and ogled,
  And his passion boil'd and bubbled,
Till he blew his silly brains out,
  And no more was by it troubled.

Charlotte, having seen his body
  Borne before her on a shutter,
Like a well-conducted person,
  Went on cutting bread and butter.

WILLIAM MAKEPEACE THACKERAY (1811–63), 'The Sorrows of Werther'

For what mad lover ever died,
To gain a soft and gentle bride?
Or for a lady tender-hearted,
In purling streams, or hemp departed?
Leap'd headlong int' Elysium
Through th' windows of a dazzling room?
But for some cross, ill-natur'd dame,
The am'rous fly burnt in his flame.

SAMUEL BUTLER (1612–80), *Hudibras*

The Abbé Rousseau was a poor young man (1784) reduced to chasing about all over the city from morning to night giving history and geography lessons. In love with one of his pupils, like Abelard with Héloïse, like Saint-Preux with Julie; less happy, certainly, but probably very near to being so; with as much passion as the latter, but more honest, more fastidious, and above all more courageous, he appears to have sacrificed himself to the object of his passion. Here is what he wrote before he blew out his brains, after having dined at a restaurant near the Palais-Royal where he betrayed no sign of anxiety or insanity; the text of the note is taken from the investigation report drawn up on the spot by the police inspector and his officers, and is sufficiently remarkable to merit preservation.

'The inexpressible contrast which exists between the nobility of my feelings and the meanness of my birth, my love for an adorable girl, as violent as it is insuperable, the fear of being the agent of her dishonour, the necessity of choosing between crime and death, all these have made me resolve to abandon life. I was born to be virtuous, I was about to be criminal; I preferred to die.'

This is a suicide worthy of admiration, and one which would be simply absurd in the moral climate of 1880.

STENDHAL, *Love*

My dear Marie, if you knew how death and love took turns crowning these last moments of my life with blossoms, those of Heaven, and those of earth, surely you would be willing to let me die. I assure you, I am wholly joyous. Mornings and evenings I kneel down, something I could never do before, and I pray to God. For this my life, the most tormented of any that anyone has ever lived, I can now at last thank Him, since He makes it good through the most glorious and sensual of deaths . . . Can it console you to hear that I would never have exchanged you for this woman if she had wanted nothing more than to live with me? . . . O I assure you, I love you so, you are so exceedingly dear and precious to me, that I can hardly say that I love this dear, this divine friend more than you. The decision that she came to in her soul, to die with me, drew me, I cannot tell you with what inexpressible and irresistible force, to her breast. Do you not recall that I many times asked you if you would die with me?—But you always answered no—A tumult of joyousness, never experienced before, gripped me, and I cannot conceal from you that her grave is more precious to me than the beds of all the empresses of this world.—Ah, my dear friend, may God soon call you to that better world where we all, loving with the love of angels, will press each other to our hearts.—Adieu.

HEINRICH VON KLEIST, to Marie von Kleist, a cousin by marriage, 21 November 1811, the day on which he shot himself and Henriette Vogel, who was incurably ill with cancer; tr. Philip B. Miller

We shall have beds round which light scents are wafted,
Divans which are as deep and wide as tombs;
Strange flowers that under brighter skies were grafted
Will scent our shelves with rare exotic blooms.

When, burning to the last their mortal ardour,
Our torch-like hearts their bannered flames unroll,
Their double light will kindle all the harder
Within the deep, twinned mirror of our soul.

One evening made of mystic rose and blue,
I will exchange a lightning-flash with you,
Like a long sob that bids a last adieu.

Later, the Angel, opening the door,
Faithful and happy, will at last renew
Dulled mirrors, and the flames that leap no more.

CHARLES BAUDELAIRE (1821–67), 'The Death of Lovers', tr. Roy Campbell

He only knew of death what all men may:
that those it takes it thrusts into dumb night.
When she herself, though,—no, not snatched away,
but tenderly unloosened from his sight,

had glided over to the unknown shades,
and when he felt that he had now resigned
the moonlight of her laughter to their glades,
and all her ways of being kind:

then all at once he came to understand
the dead through her, and joined them in their walk,
kin to them all; he let the others talk,

and paid no heed to them, and called that land
the fortunately-placed, the ever-sweet.—
And groped out all its pathways for her feet.

RILKE, 'The Death of the Beloved', tr. J. B. Leishman

Whose love is given over-well
Shall look on Helen's face in hell,
Whilst they whose love is thin and wise
May view John Knox in paradise.

DOROTHY PARKER (1893–1967), 'Partial Comfort'

Why hoard your maidenhead? There'll not be found
A lad to love you, girl, under the ground.
Love's joys are for the quick; but when we're dead
It's dust and ashes, girl, will go to bed.

ASCLEPIADES (*fl.* 290 BC), tr. R. A. Furness

Had we but world enough, and time,
This coyness, Lady, were no crime.
We would sit down, and think which way
To walk, and pass our long love's day . . .
  But at my back I always hear
Time's wingèd chariot hurrying near:
And yonder all before us lie
Deserts of vast eternity.
Thy beauty shall no more be found;
Nor, in thy marble vault, shall sound
My echoing song; then worms shall try
That long preserv'd virginity:
And your quaint honour turn to dust,
And into ashes all my lust.
The grave's a fine and private place,
But none, I think, do there embrace.

MARVELL, from 'To His Coy Mistress'

Let there be laid, when I am dead,
Ere 'neath the coffin-lid I lie,
Upon my cheek a little red,
A little black about the eye.

For I in my close bier would fain,
As on the night his vows were made,
Rose-red eternally remain,
With kohl beneath my blue eye laid.

Wind me no shroud of linen down
My body to my feet, but fold
The white folds of my muslin gown
With thirteen flounces as of old.

This shall go with me where I go:
I wore it when I won his heart;
His first look hallowed it, and so,
For him, I laid the gown apart.

No immortelles, no broidered grace
Of tears upon my cushions be;
Lay me on my pillow's lace,
My hair across it like a sea.

That pillow, those mad nights of old,
Has seen our slumbering brows unite,
And 'neath the gondola's black fold
Has counted kisses infinite.

Between my hands of ivory,
Together set for prayer and rest,
Place then the opal rosary
The holy Pope at Rome has blest.

I will lie down then on that bed
And sleep the sleep that shall not cease;
His mouth upon my mouth has said
Pater and *Ave* for my peace.

THÉOPHILE GAUTIER (1811–72), 'Posthumous Coquetry', tr. Arthur Symons

When my grave is broke up again
Some second guest to entertain
(For graves have learn'd that woman-head,
To be to more than one a bed),
  And he that digs it spies
A bracelet of bright hair about the bone,
  Will he not let'us alone,
And think that there a loving couple lies,
Who thought that this device might be some way
To make their souls, at the last busy day,
Meet at this grave, and make a little stay?

DONNE, from 'The Relic'

'Sucking the dandelion roots—'
That's a poor milk, you'll agree
When I sucked a virgin's milk
And kings knelt to me.

No, I did not 'rise again'.
After they buried me
I lay under the sand here, dry
As skulls touching the tree.

Big fires in the sky, you say,
Dry up the sea.
Act the two-backed pure beast,
Lovers, on the sand over me.

GEOFFREY GRIGSON (b. 1905), 'A Sandy Burial'

This is Anacreon's grave. Here lie
the shreds of his exuberant lust,
but hints of perfume linger by
his gravestone still, as if he must
have chosen for his last retreat
a place perpetually on heat.

ANTIPATER OF SIDON (*fl.* 120 BC), tr. Robin Skelton

Stand close around, ye Stygian set,
  With Dirce, in one boat conveyed!
Or Charon, seeing, may forget
  That he is old and she a shade.

WALTER SAVAGE LANDOR (1775–1864), 'Dirce'

Penthesileia, dead of profuse wounds,
Was despoiled of her arms by Prince Achilles,
Who, for love of that fierce white naked corpse,
Necrophily on her committed
In the public view.

Some gasped, some groaned, some bawled their indignation,
Achilles nothing cared, distraught by grief,
But suddenly caught Thersites' obscene snigger
And with one vengeful buffet to the jaw
Dashed out his life.

This was a fury few might understand,
Yet Penthesileia, hailed by Prince Achilles
On the Elysian plain, pauses to thank him
For avenging her insulted womanhood
With sacrifice.

GRAVES, 'Penthesileia'

One of the best of these [stories] is told by the surgeon Antoine Louis in a book about premature burial [1740]. As we shall see, this was not simply a case of apparent death, but of love; serious works on death are never completely free of ambiguity. A young gentleman was forced to take religious orders without vocation. While travelling he stopped at an inn. The owners were in mourning for their daughter, who had just died. 'Since the girl was not to be buried until the next day, they asked the monk to watch over her body during the night. All that he had heard about her beauty having aroused his curiosity, he uncovered the face of the would-be corpse, and far from finding her disfigured by the horrors of death, he found there certain lively graces that, causing him to forget the sanctity of his vows and stifling the lugubrious thoughts that death naturally inspires, incited him to take with the dead girl the same liberties that the sacrament would have authorized had she been alive.' The monk made love to a corpse.

But in reality, the corpse was not a corpse . . . 'The dead girl came back to life' after the monk's departure, and 'nine months later, to the great astonishment of her parents and herself, gave birth to a child. The monk happened to stop at the inn at the time and, feigning surprise upon finding alive one whom he had pretended to believe dead, acknowledged that he was the father of the child, after having been released from his vows.'

ARIÈS, *The Hour of Our Death*

There remains, however, one solid and incontestable fact which was observed in virtually all properly recorded cases. The attack on the afflicted person by a 'known vampire' usually, but not exclusively, began with kisses on the victim's throat. The kisses would eventually turn into a bite, and the gushing blood would be sucked by the vampire. There were never any complaints of pain recorded. On the contrary, most reports spoke of a kind of euphoric delirium into which the victim would slip during the blood-sucking act . . .

No one made the link between the kiss and the vampire bite more immediate and palpably real than Heinrich von Kleist, an eighteenth-century Prussian officer and dramatist. His *Penthesilea* provides a masterful portrayal of the awakening of the vampire instinct in a beautiful young woman during a moment of rare passion. Penthesilea, the Amazon queen, is in love with Achilles, and mistakenly thinks he has rejected her. In her confusion she challenges to single combat, 'with all the terrors of weaponry', the youth whom she really wanted to make love to and cherish more than anything else.

Having set the dogs on Achilles, she 'strikes her teeth into his white breast; she and her dogs—they on the right, she on the left; and . . . blood dripped from her mouth and hands.' Penthesilea's inadvertent passing from the intended love kiss to an act of savage vampirism indicates the thin dividing line between the two impulses: 'Did I kiss him to death?' she asks . . . 'Did I not kiss him? Or did I tear him to pieces? If so it was a mistake; for kissing [*Küsse*] rhymes with biting [*Bisse*], and whoever loves with her whole heart might mistake the one for the other.'

GABRIEL RONAY, *The Dracula Myth*, 1972

Here lie two poor Lovers, who had the mishap,
Though very chaste people, to die of a Clap.

POPE, 'Epitaph on the Stanton-Harcourt Lovers', killed by lightning

'Still, Dear, it is incredible to me
That here, alone,
You should have sewed him up until he died,
And in this very bed. I do not see
How you could do it, seeing what might betide.'

'Well, he came home one midnight, liquored deep—
 Worse than I'd known—
And lay down heavily, and soundly slept:
Then, desperate driven, I thought of it, to keep
Him from me when he woke. Being an adept

'With needle and thimble, as he snored, click-click
 An hour I'd sewn,
Till, had he roused, he couldn't have moved from bed,
So tightly laced in sheet and quilt and tick
He lay. And in the morning he was dead.

'Ere people came I drew the stitches out,
 And thus 'twas shown
To be a stroke.'—'It's a strange tale!' said he.
'And this same bed?'—'Yes, here it came about.'
'Well, it sounds strange—told here and now to me.

'Did you intend his death by your tight lacing?'
 'O, that I cannot own.
I could not think of else that would avail
When he should wake up, and attempt embracing.'—
 'Well, it's a cool queer tale!'

HARDY, 'Her Second Husband Hears Her Story'

Lover
Your beauty, ripe, and calm
 As Eastern Summers are,
Must now, forsaking Time and Flesh,
 Add light to some small star.

Philosopher
Whilst she yet lives, were Stars decay'd,
 Their light by hers, relief might find:
But Death will lead her to a shade
 Where Love is cold, and Beauty blind.

Lover
Lovers (whose Priests all Poets are)
 Think ev'ry Mistress, when she dies,
Is chang'd at least into a Star:
 And who dares doubt the Poets wise?

Philosopher

But ask not Bodies doom'd to die,
To what abode they go;
Since Knowledge is but Sorrow's spy
It is not safe to know.

SIR WILLIAM DAVENANT (1606–68), 'The Philosopher and the Lover: To a Mistress Dying'

Thou sallow picture of my poison'd love,
My study's ornament, thou shell of death,
Once the bright face of my betrothèd Lady . . .

And now methinks I could e'en chide myself
For doting on her beauty, though her death
Shall be reveng'd after no common action.
Does the silkworm expend her yellow labours
For thee? For thee does she undo herself?
Are lordships sold to maintain ladyships
For the poor benefit of a bewitching minute?
Why does yon fellow falsify highways
And put his life between the judge's lips
To refine such a thing? Keeps horse and men
To beat their valours for her? . . .
Who now bids twenty pound a night, prepares
Music, perfumes, and sweetmeats? All are hush'd.
Thou may'st lie chaste now!

CYRIL TOURNEUR (?1575–1626), *The Revenger's Tragedy* (Vindice addresses the skull of his mistress)

Death devours all lovely things;
Lesbia with her sparrow
Shares the darkness—presently
Every bed is narrow.

Unremembered as old rain
Dries the sheer libation,
And the little petulant hand
Is an annotation.

After all, my erstwhile dear,
My no longer cherished,
Need we say it was not love,
Now that love is perished?

EDNA ST VINCENT MILLAY (1892–1950), 'Passer Mortuus Est'

Three weeks gone and the combatants gone
returning over the nightmare ground
we found the place again, and found
the soldier sprawling in the sun.

The frowning barrel of his gun
overshadowing. As we came on
that day, he hit my tank with one
like the entry of a demon.

Look. Here in the gunpit spoil
the dishonoured picture of his girl
who has put: *Steffi. Vergissmeinnicht*
in a copybook gothic script.

We see him almost with content,
abased, and seeming to have paid
and mocked at by his own equipment
that's hard and good when he's decayed.

But she would weep to see today
how on his skin the swart flies move;
the dust upon the paper eye
and the burst stomach like a cave.

For here the lover and killer are mingled
who had one body and one heart.
And death who had the soldier singled
has done the lover mortal hurt.

KEITH DOUGLAS (1920–44), 'Vergissmeinnicht', Tunisia 1943

He turned abruptly to the fire, and continued, with what, for lack of a better word, I must call a smile—

'I'll tell you what I did yesterday! I got the sexton, who was digging Linton's grave, to remove the earth off her coffin lid, and I opened it. I thought, once, I would have stayed there: when I saw her face again—it is hers yet!—he had hard work to stir me, but he said it would change if the air blew on it, and so I struck one side of the coffin loose, and covered it up—not Linton's side, damn him! I wish he'd been soldered in lead—and I bribed the sexton to pull it away when I'm laid there, and slide mine out too, I'll have it made so; and then by the time Linton gets to us he'll not know which is which!'

'You were very wicked, Mr Heathcliff,' I exclaimed; 'were you not ashamed to disturb the dead?'

'I disturbed nobody, Nelly,' he replied; 'and I gave some ease to myself. I shall be a great deal more comfortable now; and you'll have a better chance of keeping me underground, when I get there.'

EMILY BRONTË, *Wuthering Heights*

'Is there ony room at your head, Saunders?
  Is there ony room at your feet?
Or ony room at your side, Saunders,
  Where fain, fain, I wad sleep?'

'There's nae room at my head, Marg'ret,
  There's nae room at my feet;
My bed it is fu' lowly now,
  Amang the hungry worms I sleep.

'Cauld mould is my covering now,
  But and my winding-sheet;
The dew it falls nae sooner down
  Than my resting-place is weet.

'But plait a wand o' bonny birk,
  And lay it on my breast;
And shed a tear upon my grave,
  And wish my saul gude rest.'

Then up and crew the red, red cock,
  And up and crew the gray:
''Tis time, 'tis time, my dear Marg'ret,
  That you were going away.

'And fair Marg'ret, and rare Marg'ret,
  And Marg'ret o' veritie,
Gin e'er ye love another man,
  Ne'er love him as ye did me.'

ANON., from 'Clerk Saunders'

'Is my team ploughing,
  That I was used to drive
And hear the harness jingle
  When I was man alive?'

Ay, the horses trample,
  The harness jingles now;
No change though you lie under
  The land you used to plough.

'Is football playing
  Along the river shore,
With lads to chase the leather,
  Now I stand up no more?'

Ay, the ball is flying,
  The lads play heart and soul;
The goal stands up, the keeper
  Stands up to keep the goal.

'Is my girl happy,
  That I thought hard to leave,
And has she tired of weeping
  As she lies down at eve?'

Ay, she lies down lightly,
  She lies not down to weep:
Your girl is well contented.
  Be still, my lad, and sleep.

'Is my friend hearty,
  Now I am thin and pine,
And has he found to sleep in
  A better bed than mine?'

Yes, lad, I lie easy,
  I lie as lads would choose;
I cheer a dead man's sweetheart,
  Never ask me whose.

A. E. HOUSMAN (1859–1936), *A Shropshire Lad*, XXVII

My love came back to me
Under the November tree
Shelterless and dim.
He put his hand upon my shoulder,
He did not think me strange or older,
Nor I, him.

FRANCES CORNFORD, 'All Souls' Night'

No doubt there was nothing extraordinary in the fact that Albertine's death had so little altered my preoccupations. When one's mistress is alive, a large proportion of the thoughts which form what one calls one's love comes to one during the hours when she is not by one's side. Thus one acquires the habit of having as the object of one's musings an absent person, and one who, even if she remains absent for a few hours only, during those hours is no more than a memory. Hence death does not make any great difference. When Aimé returned, I asked him to go down to Châtellerault, and thus by virtue not only of my thoughts, my sorrows, the emotion caused me by a name connected, however remotely, with a certain person, but also of all my actions, the inquiries that I undertook, the use that I made of my money, all of which was devoted to the discovery of Albertine's actions, I may say that throughout the whole of that year my life remained fully occupied with a love affair, a veritable liaison. And she who was its object was dead. It is often said that something may survive of a person after his death, if that person was an artist and put a little of himself into his work. It is perhaps in the same way that a sort of cutting taken from one person and grafted on to the heart of another continues to carry on its existence even when the person from whom it had been detached has perished.

PROUST, *Remembrance of Things Past*

That time of life thou mayst in me behold,
When yellow leaves, or none, or few do hang
Upon those boughs which shake against the cold,
Bare ruined choirs, where late the sweet birds sang.
In me thou seest the twilight of such day,
As after sunset fadeth in the west,
Which by and by black night doth take away,
Death's second self that seals up all in rest.
In me thou seest the glowing of such fire,
That on the ashes of his youth doth lie,
As the death-bed, whereon it must expire,
Consumed with that which it was nourished by.
  This thou perceiv'st, which makes thy love more strong,
  To love that well, which thou must leave ere long.

SHAKESPEARE, *Sonnets*, 73

Set me as a seal upon thine heart, as a seal upon thine arm: for love is strong as death; jealousy is cruel as the grave: the coals thereof are coals of fire, which hath a most vehement flame.

Many waters cannot quench love, neither can the floods drown it: if a man would give all the substance of his house for love, it would utterly be contemned.

Song of Solomon, 8

But true love is a durable fire,
  In the mind ever burning,
Never sick, never old, never dead,
  From itself never turning.

SIR WALTER RALEGH (*c.*1552–1618), from 'As you came from the holy land'

Children

THIS section was by far the most difficult to cope with. To face the death of a child with anything resembling equanimity requires impregnable religious convictions and a firm degree of contempt for this world and this life. ('Yet even so,' the Buddhist poet Issa admitted in his grief, 'yet even so'.) Our earthly existence has to be seen as no more than a bridge between two heavens, if we are lucky—'not an inn, but an hospital'—to be crossed as decently as possible; and those who die young make the passage expeditiously and assured of guiltlessness, possessors of the kingdom of heaven, not merely inheritors, as Coleridge has it. Or, in the secular terms of the Greek poet: though the young have experienced few joys, at least they have suffered few sorrows.

For many the death of a child will have been their equivalent of Voltaire's Lisbon earthquake:

> Say, will you then eternal laws maintain,
> Which God to cruelties like these constrain?
> Whilst you these facts replete with horror view,
> Will you maintain death to their crimes was due?
> And can you then impute a sinful deed
> To babes who on their mothers' bosoms bleed?

If from such afflictions you might infer the necessary existence of a heaven in which all is more than put right, you could as well deduce the necessary non-existence of a God worth the name. In these circumstances more than any others Christians have to fall back on the mysteriousness of the workings of Providence.

There could scarcely be any humour, any light relief, here were it not for Freud's view of children, in which he sounds much like the petulant narrator of 'We Are Seven', and more especially children's views of death. Children have to learn about death, and the processes of learning often provide the onlooker with amusement. Not that Derwent May's poem does exactly that; it bears incidentally on one species of limited immortality, won through old films shown on television. The fears and anxieties of children, since they are not unlike those of grown-ups, are less amusing to the onlooker. And at times children can appear more adult than the grown-ups who seek to soften for them the blows of knowledge and the facts of life.

The risk of what in other spheres we would more readily condemn as sentimentality is strong, of course. Whatever brings a measure of solace—and Dickens must surely have done this—is not to be sneered at. That I have not included the death of Little Nell does not imply agreement with the clever, heartless epigram attributed to Oscar Wilde. I was tempted to cite the latter's own 'Requiescat'—

> Lily-white, white as snow,
> She hardly knew
> She was a woman, so
> Sweetly she grew

—if only to suggest that in the matter of sentimentality we like to gulp down our cake as well as spurn it. But perhaps to anthologize on this subject does require a heart of stone.

~

The fear of death has no meaning to a child; hence it is that he will play with the dreadful word and use it as a threat against a playmate: 'If you do that again, you'll die, like Franz!' Meanwhile the poor mother gives a shudder and remembers, perhaps, that the greater half of the human race fail to survive their childhood years. It was actually possible for a child, who was over eight years old at the time, coming home from a visit to the Natural History Museum, to say to his mother: 'I'm so fond of you, Mummy: when you die, I'll have you stuffed and I'll keep you in this room, so that I can see you *all* the time.' So little resemblance is there between a child's idea of being dead and our own!

Footnote added 1909: I was astonished to hear a highly intelligent boy of ten remark after the sudden death of his father: 'I know father's dead, but what I can't understand is why he doesn't come home to supper.'

FREUD, *The Interpretation of Dreams*, tr. James Strachey

> —A simple Child,
> That lightly draws its breath,
> And feels its life in every limb,
> What should it know of death?

I met a little cottage Girl:
She was eight years old, she said;
Her hair was thick with many a curl
That clustered round her head . . .

'Sisters and brothers, little Maid,
How many may you be?'
'How many? Seven in all,' she said,
And wondering looked at me.

'And where are they? I pray you tell.'
She answered, 'Seven are we;
And two of us at Conway dwell,
And two are gone to sea.

Two of us in the churchyard lie,
My sister and my brother;
And, in the churchyard cottage, I
Dwell near them with my mother.'

'You say that two at Conway dwell,
And two are gone to sea,
Yet ye are seven! I pray you tell,
Sweet Maid, how this may be.'

Then did the little Maid reply,
'Seven boys and girls are we;
Two of us in the churchyard lie,
Beneath the churchyard tree.'

'You run about, my little Maid,
Your limbs they are alive;
If two are in the churchyard laid,
Then ye are only five.'

'Their graves are green, they may be seen,'
The little Maid replied,
'Twelve steps or more from my mother's door,
And they are side by side . . .

The first that died was sister Jane;
In bed she moaning lay,
Till God released her of her pain
And then she went away.

So in the churchyard she was laid;
And, when the grass was dry,
Together round her grave we played,
My brother John and I.

And when the ground was white with snow,
And I could run and slide,
My brother John was forced to go,
And he lies by her side.'

'How many are you, then,' said I,
'If they two are in heaven?'
Quick was the little Maid's reply,
'O Master! we are seven.'

'But they are dead; those two are dead!
Their spirits are in heaven!'
'Twas throwing words away; for still
The little Maid would have her will,
And said, 'Nay, we are seven!'

WORDSWORTH, from 'We Are Seven'

Mother, mother, I feel sick,
Send for the doctor, quick, quick, quick.
Doctor, doctor, shall I die?
Yes, my dear, and so shall I.
How many carriages shall I have?
One, two, three, four . . .

Skipping rhyme

The dream illustrates subconscious preoccupation with the idea of death in a young child, for it came repeatedly at the age of six. Night after night my sister's high chair turned into a devil, chased me across the nursery floor, and caught me. At that I died—and awoke shaking with a terror which I kept to myself. No good waking Nanny to tell her that one had died. She would only say, 'Nonsense, child! Shut up and go to sleep.' How many children do the same?

ROSALIND HEYWOOD, *Man's Concern With Death*

I saw death. At the age of five: it was watching me; in the evenings, it prowled on the balcony: it pressed its nose to the window; I used to see it but I did not dare to say anything. Once, on the Quai Voltaire, we met it: it was a tall, mad old woman, dressed in black, who mumbled as she went by: 'I shall put that child in my pocket.' Another time, it took the form of a hole: this was at Arcachon; [we] were visiting Madame Dupont and her son Gabriel, the composer. I was playing in the garden of the villa, scared because I had been told that Gabriel was ill and was going to die. I was playing at horses, half-heartedly, and galloping round the house. Suddenly, I noticed a gloomy hole: the cellar, which had been opened; an indescribable impression of loneliness and horror blinded me: I turned round and, singing at the top of my voice, I fled. At that time, I had an assignation with it every night in my bed. It was a ritual: I had to sleep on my left side, my face to the wall; I would wait, trembling all over, and it would appear, a very conventional skeleton, with a scythe; I then had permission to turn on my right side, it would go away and I could sleep in peace.

JEAN-PAUL SARTRE (1905–80), *Words*, tr. Irene Clephane

With this desire of physical beauty mingled itself early the fear of death—the fear of death intensified by the desire of beauty . . . The child had heard indeed of the death of his father, and how, in the Indian station, a fever had taken him, so that though not in action he had yet died as a soldier; and hearing of the 'resurrection of the just', he could think of him as still abroad in the world, somehow, for his protection—a grand, though perhaps rather terrible figure, in beautiful soldier's things, like the figure in the picture of Joshua's Vision in the Bible—and of that, round which the mourners moved so softly, and afterwards with such solemn singing, as but a worn-out garment left at a deserted lodging. So it was, until on a summer day he walked with his mother through a fair churchyard. In a bright dress he rambled among the graves, in the gay weather, and so came, in one corner, upon an open grave for a child—a dark space on the brilliant grass—the black mould lying heaped up round it, weighing down the little jewelled branches of the dwarf rose bushes in flower. And therewith came, full-grown, never wholly to leave him, with the certainty that even children do sometimes die, the physical horror of death, with its wholly selfish recoil from the association of lower forms of life, and the suffocating weight above. No benign, grave figure in beautiful soldier's things any longer abroad in the world for his protection! only a few poor, piteous bones; and above them, possibly, a certain sort of figure he hoped not to see. For sitting one day in the garden below an open window, he heard people talking, and could not but listen, how, in a sleepless hour, a sick woman had seen one of the dead sitting beside her, come to call her hence; and from the broken talk evolved with much clearness the notion that not all those dead people had really departed to the churchyard, nor were quite so motionless as they looked, but led a secret, half-fugitive life in their old homes, quite free by night, though sometimes visible in the day, dodging from room to room, with no great goodwill towards those who shared the place with them. All night the figure sat beside him in the reveries of his broken sleep, and was not quite gone in the morning—an odd, irreconcilable new member of the household, making the sweet familiar chambers unfriendly and suspect by its uncertain presence. He could have hated the dead he had pitied so, for being thus. Afterwards he came to think of those poor, home-returning ghosts, which all men have fancied to themselves—the *revenants*—pathetically, as crying, or beating with vain hands at the doors, as the wind came, their cries distinguishable in it as a wilder inner note. But, always making death more unfamiliar still, that old experience would ever, from time to time, return to him; even in the living he sometimes caught its likeness; at any time or place, in a moment, the faint atmosphere of the chamber of death would be breathed around him, and the image with the bound chin, the quaint smile, the straight, stiff feet, shed itself across the air upon the bright carpet, amid the gayest company, or happiest communing with himself.

WALTER PATER (1839–94), 'The Child in the House'

'My father is dead?' I said. 'What does that mean?'
'That means you won't see him again.'
'But how? I won't see Father again?'
'No.'
'And why shan't I see him again?'
'Because God has taken him back from you.'
'Forever?'
'Forever.'
'And you say I shan't see him again?'
'Never again.'
'Never again, never?'
'Never!'

'And where does God live?'

'He lives in the sky.'

I remained thoughtful for a while. Though such a child, and unable to reason, I understood nevertheless that something final had happened in my life. Then, seizing the first moment when nobody was paying attention to me, I escaped from my uncle's house and ran straight to my mother's.

All the doors were open, all the faces showed distress. Death could be sensed there.

I entered without being noticed. I reached a small room where arms were kept; I took down a single-barrelled gun which belonged to my father, and which had often been promised me when I grew up.

Then, armed with the gun, I went upstairs.

On the first-floor landing I met my mother. She was coming out of the death-chamber . . . she was in tears.

'Where are you going?' she asked, astonished to see me there when she thought I was with my uncle.

'I'm going to the sky!' I answered.

'What? You're going to the sky?'

'Yes, don't stop me.'

'And what are you going to do in the sky, my poor child?'

'I'm going to kill God, who killed Father.'

My mother clutched me in her arms, squeezing enough to suffocate me.

'Oh, don't say such things, my child,' she cried. 'We're quite unhappy enough already!'

ALEXANDRE DUMAS (1802–70), *My Memoirs* (his father, General Dumas, died in 1806)

'At home,' said Mr Facey, in a low impressive tone, 'there be a large black book, and in 'en be written they same words, but there be more of them. "Thou shalt not steal" be written and close along be written, too, " 'Tain't no stealing to take for a dumb animal." '

'Oh,' said Lily, 'and who put that in?'

'God Himself,' replied Mr Facey, 'did write they words, for ink be faded and washed by the flood waters. Second words be always best, too.'

'Not when our Daddy do talk,' said Lily.

'But they be best,' continued Mr Facey, 'when God be the writer, for 'e did mind they dumb things and did put in they new words.'

Lily stood upon the log and gave her hair a little shake.

'What be a dumb animal, Mr Facey?' she inquired.

'Whatever,' replied Mr Facey, 'do bide about and say nothing.'

Lily pondered over his words . . .

Lily Topp awoke the next morning feeling hungry. She also awoke surprised, for her cot had been moved in the night, and she with it, into a small inner room.

Lily was soon in her clothes and, hearing the cottage door open and shut, she peeped from the window and saw her mother walking up the lane with a hurried step and crying as she walked.

'She bain't dumb,' said Lily, a little disappointedly.

Lily's room faced the meadow—the snared rabbit still remained there; nothing had touched it.

Lily went into the next room. Her father was in bed, but a sheet was drawn over his face.

Lily pulled the sheet back.

'Bain't 'ee going to get up, our Dad?' she called.

There was no reply and Lily became interested.

'Be thee dumb?' asked Lily. But no reply came.

'If thee be a dumb animal,' said Lily, 'who do bide about and say nothing, thee may be fed.' John Topp remained silent.

Lily went at once into the field in front of the house . . .

Farmer Denny always rose early, as a rich man should do, in order to guard his wealth from harm, and as Lily was loosing the noose from the dead rabbit's neck the heavy hand of the farmer was laid upon her shoulder.

Lily shook herself free and stood up boldly with the rabbit in her hand.

''Tain't stealing,' she said, 'to take for a dumb animal.'

'And what dumb animal are you taking my rabbit for?' asked Mr Denny.

'Our Daddy do bide about and say nothing,' replied Lily.

Farmer Denny moved out of the child's way. He had heard a sound that pleaded for the dumb animal, too—the tolling of the parish bell.

T. F. POWYS (1875–1953), 'A Dumb Animal'

About four weeks after mother had died, Wendy (aged four years) complained that no one loved her. In an attempt to reassure her, father named a long list of people who did . . . On this Wendy commented aptly, 'But when my mommy wasn't dead I didn't need so many people—I needed just one.' Four months after mother's death, when the family took a spring vacation in Florida, it was evident that Wendy's forlorn hopes of mother's return persisted. Repeating as it did an exceptionally enjoyable holiday there with mother the previous spring, Wendy was enthusiastic at the prospect and during the journey recalled with photographic accuracy every incident of the earlier one. But after arrival she was whiny, complaining and petulant. Father talked with her about the sad and happy memories the trip evoked and how very tragic it was for all of them that Mommy would never return; to which Wendy responded wistfully, 'Can't Mommy move in the grave just a little bit?' Wendy's increasing ability to come to terms with the condition of dead people was expressed a year after mother's death when a distant relative died. In telling Wendy about it father, eager not to upset her, added that the relative would be comfortable in the ground because he would be protected by a box. Wendy replied, 'But if he's really dead, why does he have to be comfortable?'

Recorded by M. J. Barnes, quoted in John Bowlby, *Loss: Sadness and Depression*, 1980

His mother reports that Stephen (four years, ten months), 'at present thinks that we all turn into statues when we die, owing to the fact that he first met Queen Victoria as a statue in Kensington Gardens and then was told that she had been dead some time.'

Ben (three years, six months) had been told that his mother would die when she was old, perhaps when over seventy. He began worrying about the end or endlessness of numbers, and was told that there was an imaginary end called infinity. The following morning, babbling in his bed before rising, he was heard to say, '*finity!*' and then, after a pause, 'Finity-one, finity-two . . .' and so on.

SYLVIA ANTHONY, *The Discovery of Death in Childhood and After*, 1971

Beautifully Janet slept
Till it was deeply morning. She woke then
And thought about her dainty-feathered hen,
To see how it had kept.

One kiss she gave her mother,
Only a small one gave she to her daddy
Who would have kissed each curl of his shining baby;
No kiss at all for her brother.

'Old Chucky, Old Chucky!' she cried,
Running on little pink feet upon the grass
To Chucky's house, and listening. But alas,
Her Chucky had died.

It was a transmogrifying bee
Came droning down on Chucky's old bald head
And sat and put the poison. It scarcely bled,
But how exceedingly

And purply did the knot
Swell with the venom and communicate
Its rigor! Now the poor comb stood up straight
But Chucky did not.

So there was Janet
Kneeling on the wet grass, crying her brown hen
(Translated far beyond the daughters of men)
To rise and walk upon it.

And weeping fast as she had breath
Janet implored us, 'Wake her from her sleep!'
And would not be instructed in how deep
Was the forgetful kingdom of death.

John Crowe Ransom (1888–1974), 'Janet Waking'

Pat came swinging along; in his hand he held a little tomahawk that winked in the sun.

'Come with me,' he said to the children, 'and I'll show you how the kings of Ireland chop the head off a duck.' . . .

There was an old stump beside the door of the fowl-house. Pat grabbed the duck by the legs, laid it flat across the stump, and almost at the same moment down came the little tomahawk and the duck's head flew off the stump. Up the blood spurted over the white feathers and over his hand.

When the children saw the blood they were frightened no longer. They crowded round him and began to scream. Even Isabel leaped about crying: 'The blood! The blood!' Pip . . . shouted, 'I saw it. I saw it', and jumped round the wood block.

Rags, with cheeks as white as paper, ran up to the little head, put out a finger as if he wanted to touch it, shrank back again and then again put out a finger. He was shivering all over.

Even Lottie, frightened little Lottie, began to laugh and pointed at the duck and shrieked: 'Look, Kezia, look.'

'Watch it!' shouted Pat. He put down the body and it began to waddle—with only a long spurt of blood where the head had been; it began to pad away without a sound towards the steep bank that led to the stream . . . That was the crowning wonder.

'Do you see that? Do you see that?' yelled Pip. He ran among the little girls tugging at their pinafores.

'It's like a little engine. It's like a funny little railway engine,' squealed Isabel.

But Kezia suddenly rushed at Pat and flung her arms round his legs and butted her head as hard as she could against his knees.

'Put head back! Put head back!' she screamed.

When he stooped to move her she would not let go or take her head away. She held on as hard as she could and sobbed: 'Head back! Head back!' until it sounded like a loud strange hiccup.

'It's stopped. It's tumbled over. It's dead,' said Pip . . .

The children stopped screaming as suddenly as they had begun. They stood round the dead duck. Rags was not frightened of the head any more. He knelt down and stroked it, now.

'I don't think the head is quite dead yet,' he said. 'Do you think it would keep alive if I gave it something to drink?'

But Pip got very cross: 'Bah! You baby.' He whistled to Snooker and went off . . .

'There now,' said Pat to Kezia. 'There's the grand little girl.'

She put up her hands and touched his ears. She felt something. Slowly she raised her quivering face and looked. Pat wore little round gold ear-rings. She never knew that men wore ear-rings. She was very much surprised.

'Do they come on and off?' she asked huskily.

KATHERINE MANSFIELD (1888–1923), 'Prelude'

Rattle creak very white
Shining and bright
Patterns in the bones
The teeth come easily

Out of the jaw-bone
Green at bottom
Dark green from
Chewing grass.
Small animals live in it now
In the little holes
Just think. When it was alive
Eyes in eye sockets
And all joined together.

LUCY JANE SIMPSON, aged 7, 'Sheep's Skull', 1970

'Daddy, how old is Groucho Marx?'
'Sorry, dear boy, he's dead.'
'Gosh! And Chico? Oh yes, and Harpo?'
'Dead. All of them dead.'
'Daddy, is Lassie very old?'
'Dogs die young, you know.'
'Will Hay's good! Is he dead too?'
'Thirty years ago.'

'Daddy, if Elvis comes this way
Can we go and hear him?'
'Elvis stays in Memphis now,
Blue carnations near him.'
'Sossidge is on again tonight.'
'That was Joyce Grenfell, eh?'
'Was? Oh, Daddy, did she die?'
'Just the other day.'

This is immortality
Never dreamed of yet:
Life because a child sits by
A television set.
'Gary Cooper's good on horses.'
'That was his last ride.'
'Disney must be very rich.'
'Was, until he died.'

But the child who's sitting there
Starts to love each day
People who at natural breaks
Death will take away.

'John Wayne—Bogey—Errol Flynn—
   Are they full of lead?'
'Darling, it wasn't quite like that—
   But all of them *are* dead.'

DERWENT MAY (b. 1930), 'A Child in the 80s'

Márgarét, áre you gríeving
Over Goldengrove unleaving?
Leáves, líke the things of man, you
With your fresh thoughts care for, can you?
Áh! ás the heart grows older
It will come to such sights colder
By and by, nor spare a sigh
Though worlds of wanwood leafmeal lie;
And yet you *will* weep and know why.
Now no matter, child, the name:
Sórrow's spríngs áre the same.
Nor mouth had, no nor mind, expressed
What heart heard of, ghost guessed:
It ís the blight man was born for,
It is Margaret you mourn for.

GERARD MANLEY HOPKINS (1844–89), 'Spring and Fall: to a young child'

They laid her in the grave—the sweet mother with her baby in her arms—while the Christmas snow lay thick upon the graves. It was Mr Cleves who buried her. On the first news of Mr Barton's calamity, he had ridden over from Tripplegate to beg that he might be made of some use, and his silent grasp of Amos's hand had penetrated like the painful thrill of life-recovering warmth to the poor benumbed heart of the stricken man.

The snow lay thick upon the graves, and the day was cold and dreary; but there was many a sad eye watching that black procession as it passed from the vicarage to the church, and from the church to the open grave. There were men and women standing in that churchyard who had bandied vulgar jests about their pastor, and who had lightly charged him with sin; but now, when they saw him following the coffin, pale and haggard, he was consecrated anew by his great sorrow, and they looked at him with respectful pity.

All the children were there, for Amos had willed it so, thinking that some dim memory of that sacred moment might remain even with little Walter, and link itself with what he would hear of his sweet mother in after years. He himself led Patty and Dickey; then came Sophy and Fred; Mr Brand had begged to carry Chubby, and Nanny followed with Walter. They made a circle round the grave while the coffin was being lowered. Patty alone of all the children felt that mamma was in that coffin, and that a new and sadder life had begun for papa and herself. She was pale and trembling, but she clasped his hand more firmly as the coffin went down, and gave no sob. Fred and Sophy, though they were only two and three years younger, and though they had seen mamma in her coffin, seemed to themselves to be looking at some strange show. They had not learned to decipher that terrible handwriting of human destiny, illness and death. Dickey had rebelled against his black clothes, until he was told that it would be naughty to mamma not to put them on, when he at once submitted; and now, though he had heard Nanny say that mamma was in heaven, he had a vague notion that she would come home again tomorrow, and say he had been a good boy and let him empty her work-box. He stood close to his father, with great rosy cheeks, and wide open blue eyes, looking first up at Mr Cleves and then down at the coffin, and thinking he and Chubby would play at that when they got home.

GEORGE ELIOT, 'Amos Barton', *Scenes of Clerical Life*

The sky was serene and bright, the air clear, perfumed with the fresh scent of newly-fallen leaves, and grateful to every sense. The neighbouring stream sparkled, and rolled onward with a tuneful sound; the dew glistened on the green mounds, like tears shed by Good Spirits over the dead.

Some young children sported among the tombs, and hid from each other, with laughing faces. They had an infant with them, and had laid it down asleep upon a child's grave, in a little bed of leaves. It was a new grave—the resting-place, perhaps, of some little creature, who, meek and patient in its illness, had often sat and watched them, and now seemed to their minds scarcely changed.

She drew near and asked one of them whose grave it was. The child answered that that was not its name; it was a garden—his brother's. It was greener, he said, than all the other gardens, and the birds loved it better because he had been used to feed them. When he had done speaking, he looked at her with a smile, and kneeling down and nestling for a moment with his cheek against the turf, bounded merrily away.

DICKENS, *The Old Curiosity Shop*

Gontran falls upon his knees at the foot of the bed; he buries his head in the sheets, but he cannot succeed in weeping. No emotion stirs his heart; his eyes remain despairingly dry. Then he gets up and looks at the impassive face on the bed. At this solemn moment, he would like to have some rare, sublime experience—hear a message from the world beyond—send his thought flying into ethereal regions, inaccessible to mortal senses. But no! his thought remains obstinately grovelling on the earth; he looks at the dead man's bloodless hands and wonders for how much longer the nails will go on growing. The sight of the unclasped hands grates on him. He would like to join them, to make them hold the crucifix. What a good idea! He thinks of Séraphine's astonishment when she sees the dead hands folded together; the thought of Séraphine's astonishment amuses him; and then he despises himself for being amused. Nevertheless he stoops over the bed. He seizes the arm which is furthest from him. The arm is stiff and will not bend. Gontran tries to force it, but the whole body moves with it. He seizes the other arm, which seems a little less rigid. Gontran almost succeeds in putting the hand in the proper place. He takes the crucifix and tries to slip it between the fingers and the thumb, but the contact of the cold flesh turns him sick . . . He gives up everything—the crucifix, which drops aslant on the tumbled sheet, and the lifeless arm, which falls back into its first position; then, through the depths of the funereal silence, he suddenly hears a rough and brutal 'God damn!' which fills him with terror, as if someone else . . . He turns round—but no! he is alone. It was from his own lips, from his own heart, that that resounding curse broke forth—his, who until today has never uttered an oath! Then he sits down and plunges again into his reading.

André Gide (1869–1951), *The Coiners*, tr. Dorothy Bussy

Now I lay me down to sleep;
I pray the Lord my soul to keep.
If I should die before I wake,
I pray the Lord my soul to take.

New England Primer, 1781

Every time an earth mother smiles over the birth of a child, a spirit mother weeps over the loss of a child.

Ashanti saying

Here she lies, a pretty bud,
Lately made of flesh and blood:
Who as soon fell fast asleep
As her little eyes did peep.
Give her strewings, but not stir
The earth that lightly covers her.

HERRICK, 'Epitaph upon a Child that died'

Thou Mother dear and thou my Father's shade,
To you I now commit the gentle maid,
Erotion, my little love, my sweet;
Let not her shuddering spirit fear to meet
The ghosts, but soothe her lest she be afraid.
How should a baby heart be undismayed
To pass the lair where Cerberus is laid?
The little six-year maiden gently greet.
Dear reverend spirits, give her kindly aid
And let her play in some Elysian glade,
Lisping my name sometimes—and I entreat
Lie softly on her, kindly earth; her feet,
Such tiny feet, on thee were lightly laid.

MARTIAL (*c*.40–*c*.104), 'For Erotion's Grave', tr. F. A. Wright

Bewail not much, my parents! me, the prey
Of ruthless Hades, and sepulchred here.
An infant, in my fifth scarce finished year,
He found all sportive, innocent, and gay,
Your young Callimachus; and if I knew
Not many joys, my griefs also were few.

LUCIAN (*c*.115–*c*.180), tr. William Cowper

Weep with me all you that read
This little story,
And know, for whom a tear you shed,
Death's self is sorry.

'Twas a child that so did thrive
In grace and feature,
As heaven and nature seemed to strive
Which owned the creature.
Years he numbered scarce thirteen
When fates turned cruel,
Yet three filled zodiacs had he been
The stage's jewel,
And did act (what now we moan)
Old men so duly
As, sooth, the Parcae thought him one,
He played so truly.
So, by error, to his fate
They all consented,
But viewing him since (alas, too late)
They have repented;
And have sought, to give new birth,
In baths to steep him;
But being so much too good for earth,
Heaven vows to keep him.

JONSON, 'Epitaph on Salomon Pavy, a Child of Queen Elizabeth's Chapel' (a boy actor who died in 1603)

'Be, rather than be called, a child of God,'
Death whispered! With assenting nod,
Its head upon its mother's breast,
The Baby bowed, without demur—
Of the kingdom of the Blest
Possessor, not Inheritor.

SAMUEL TAYLOR COLERIDGE (1772–1834), 'On an Infant which died before Baptism'

Farewell, thou child of my right hand, and joy;
My sin was too much hope of thee, lov'd boy,
Seven years th'wert lent to me, and I thee pay,
Exacted by thy fate, on the just day.
O, could I lose all father now. For why
Will man lament the state he should envy?

To have so soon 'scaped world's, and flesh's rage,
And, if no other misery, yet age?
Rest in soft peace, and, ask'd, say here doth lie
BEN JONSON his best piece of poetry.
For whose sake, henceforth, all his vows be such,
As what he loves may never like too much.

JONSON, 'On My First Son'

Yesterday morning William walked as far as the Swan with Aggy Fisher. She was going to attend upon Goan's dying Infant. She said, 'There are many heavier crosses than the death of an Infant', and went on, 'There was a woman in this vale who buried 4 grown-up children in one year, and I have heard her say when many years were gone by that she had more pleasure in thinking of those 4 than of her living Children, for as Children get up and have families of their own their duty to their parents "*wears out and weakens*". She could trip lightly by the graves of those who died when they were young, with a light step, as she went to Church on a Sunday.'

DOROTHY WORDSWORTH, *Grasmere Journals*, 3 June 1802

'Now lay me down,' he said, 'and, Floy, come close to me, and let me see you!'

Sister and brother wound their arms around each other, and the golden light came streaming in, and fell upon them, locked together.

'How fast the river runs, between its green banks and the rushes, Floy! But it's very near the sea. I hear the waves! They always said so!'

Presently he told her that the motion of the boat upon the stream was lulling him to rest. How green the banks were now, how bright the flowers growing on them, and how tall the rushes! Now the boat was out at sea, but gliding smoothly on. And now there was a shore before him. Who stood on the bank!—

He put his hands together, as he had been used to do at his prayers. He did not remove his arms to do it; but they saw him fold them so, behind her neck.

'Mama is like you, Floy. I know her by the face! But tell them that the print upon the stairs at school is not divine enough. The light about the head is shining on me as I go!'

The golden ripple on the wall came back again, and nothing else stirred in the room. The old, old fashion! The fashion that came in with our first garments, and will last unchanged until our race has run its course, and the wide firmament is rolled up like a scroll. The old, old fashion—Death!

Oh thank God, all who see it, for that older fashion yet, of Immortality! And look upon us, angels of young children, with regards not quite estranged, when the swift river bears us to the ocean!

DICKENS, *Dombey and Son* (the death of Paul Dombey)

[1658, 27 January] . . . so as all artificial help failing, & his natural strength exhausted, we lost the prettiest, and dearest Child, that ever parents had, being but 5 years & 3 days old in years but even at that tender age, a prodigy for Wit, & understanding; for beauty of body a very Angel, & for endowments of mind, of incredible & rare hopes. To give only a little taste of some of them, & thereby glory to God (who out of the mouths of Babes & Infants does sometimes perfect his praises), at 2 years & half old he could perfectly read any of the English, Latin, French or Gothic letters, pronouncing the three first languages exactly. He had before the 5th year or in that year . . . begun himself to write legibly, & had a strange passion for Greek: the number of Verses he could recite was prodigious, & what he remembered of the parts of plays, which he would also act: & when seeing a Plautus in one's hand, he asked what book it was, & being told it was Comedy &c, & too difficult for him, he wept for sorrow . . . When one told him how many days a certain Quaker had fasted in Colchester, he replied, that was no wonder; for Christ had said, That Man should not live by bread alone, but by the word of God . . . The day before he died, he called to me, & in a more serious manner than usually, told me, That for all I loved him so dearly, I would give my house, land & all my fine things to his Bro. Jack, he should have none of them, & next morning when first he found himself ill, & that I persuaded him to keep his hands in bed, he demanded, whether he might pray to God with his hands unjoined, & a little after, whilst in great agony, whether he should not offend God, by using his holy name so oft, calling for Ease . . . but thus God having dressed up a Saint fit for himself, would not permit him longer with us, unworthy of the future fruits of this incomparable hopeful blossom: such a Child I never saw . . .

JOHN EVELYN, *Diary*

Grief fills the room up of my absent child,
Lies in his bed, walks up and down with me,
Puts on his pretty looks, repeats his words,
Remembers me of all his gracious parts,
Stuffs out his vacant garments with his form:
Then have I reason to be fond of grief.

SHAKESPEARE, *King John*

The little cousin is dead, by foul subtraction,
A green bough from Virginia's aged tree,
And none of the county kin like the transaction,
Nor some of the world of outer dark, like me.

A boy not beautiful, nor good, nor clever,
A black cloud full of storms too hot for keeping,
A sword beneath his mother's heart—yet never
Woman bewept her babe as this is weeping.

A pig with a pasty face, so I had said,
Squealing for cookies, kinned by poor pretence
With a noble house. But the little man quite dead,
I see the forebears' antique lineaments.

The elder men have strode by the box of death
To the wide flag porch, and muttering low send round
The bruit of the day. O friendly waste of breath!
Their hearts are hurt with a deep dynastic wound.

He was pale and little, the foolish neighbours say;
The first-fruits, saith the Preacher, the Lord hath taken;
But this was the old tree's late branch wrenched away,
Grieving the sapless limbs, the shorn and shaken.

RANSOM, 'Dead Boy'

They look up with their pale and sunken faces,
    And their looks are sad to see,
For the man's hoary anguish draws and presses
    Down the cheeks of infancy—
'Your old earth,' they say, 'is very dreary';
    'Our young feet,' they say, 'are very weak!
Few paces have we taken, yet are weary—
    Our grave-rest is very far to seek.

Ask the aged why they weep, and not the children,
  For the outside earth is cold—
And we young ones stand without, in our bewildering,
  And the graves are for the old.'

'True,' say the young children, 'it may happen
  That we die before our time:
Little Alice died last year—the grave is shapen
  Like a snowball, in the rime.
We looked into the pit prepared to take her—
  Was no room for any work in the close clay:
From the sleep wherein she lieth none will wake her,
  Crying, "Get up, little Alice! it is day."
If you listen by that grave, in sun and shower,
  With your ear down, little Alice never cries!—
Could we see her face, be sure we should not know her,
  For the smile has time for growing in her eyes—
And merry go her moments, lulled and stilled in
  The shroud, by the kirk-chime!
It is good when it happens,' say the children,
  'That we die before our time.' . . .

'For oh,' say the children, 'we are weary,
  And we cannot run or leap—
If we cared for any meadows, it were merely
  To drop down in them and sleep.
Our knees tremble sorely in the stooping—
  We fall upon our faces, trying to go;
And, underneath our heavy eyelids drooping,
  The reddest flower would look as pale as snow.
For, all day, we drag our burden tiring,
  Through the coal-dark, underground—
Or, all day, we drive the wheels of iron
  In the factories, round and round.'

ELIZABETH BARRETT BROWNING (1806–61), from 'The Cry of the Children'

At about 1.30 a.m. a thirteen-year-old boy was found suffering in front of Nakanoshima Public Hall by a policeman. The youth was treated at a hospital but died at 2 p.m. the same day. He had taken rat poison.

Just before his death he faintly related the tragic story of his short life. He gave his name as Kazuo Yamamoto. He became an orphan through war ravage and was sent to an orphanage in Tokyo. Leaving the orphanage, he had been earning his living shining shoes before coming to Osaka.

His last words were: 'I wanted to die because of a headache.' He had no personal belongings with him.

Japanese news item, 1954

Is it children we are killing now? My God,
What are we? Savages? Just let me ask you,
This morning he was playing, by the window—
To think they've killed the poor little mite!
He was out in the street, and they shot him there.
He was kind and sweet, Monsieur, like Jesus.
Me, I'm old, it's fair enough for me to go—
It wouldn't have hurt this Monsieur Bonaparte
To take my life instead of this poor child's! . . .
I had nothing left of his mother but him.
Why was he killed? If only someone could explain.
He'd never shouted 'Vive la République' . . .

VICTOR HUGO (1802–85), from 'A Recollection of the Night of the Fourth'

It would be impossible to say what horrors were embedded in the minds of the children who lived through the day of the bombing of Hiroshima. On the surface their recollections, months after the disaster, were of an exhilarating adventure. Toshio Nakamura, who was ten at the time of the bombing, was soon able to talk freely, even gaily, about the experience, and a few weeks before the anniversary he wrote the following matter-of-fact essay for his teacher at Noboricho Primary School: 'The day before the bomb, I went for a swim. In the morning, I was eating peanuts. I saw a light. I was knocked to little sister's sleeping place. When we were saved I could only see as far as the tram. My mother and I started to pack our things. The neighbours were walking around burned and bleeding. Hataya-san told me to run away with her. I said I wanted to wait for my mother. We went to the park. A whirlwind came. At night a gas tank burned and I saw the reflection in the river. We stayed in the park one night. Next day I went to Taiko Bridge and met my girl friends Kikuki and Murakami. They were looking for their mothers. But Kikuki's mother was wounded and Murakami's mother, alas, was dead.'

JOHN HERSEY, *Hiroshima*, 1946

Who is it,
Tapping on the back door?
Tap, tap, tap . . .
Who is it,
Shaking the door?
Your mother is working in the house,
Your mother is waiting for you,
Your mother still has your pants and your shirts.

KAZUKO YAMADA, 'The Wind', from *The Songs of Hiroshima*, 1955, tr. Miyao Ohara and D. J. Enright

The children cried: 'Mummy!'
'I have been good!'
'Why is it dark! Dark!'

You can see them
going down
you can see the marks
of small feet here and there
going down

Their pockets full
of string and pebbles
and little horses made of wire

The great plain closed
like a geometric figure
one tree of black smoke
vertical
a dead tree
starless its crown

TADEUSZ RÓŻEWICZ (b. 1921), 'Massacre of the innocents', tr. Jan Darowski

They shall hunger no more, neither thirst any more; neither shall the sun light on them, nor any heat. For the Lamb which is in the midst of the throne shall feed them, and shall lead them unto living fountains of water: and God shall wipe away all tears from their eyes.

The Order for the Burial of a Child; Revelation, 7

Another region of the beyond was called *Chichihuacuauhco*. The name is composed of *chichihua*, 'wet-nurse', *cuáuhitl*, 'tree', and *co*, 'place', so that the word means 'in the wet-nurse tree'. To this place went the children who died before attaining the age of reason. There they were nourished by the milk which fell in drops from the tree.

MIGUEL LÉON-PORTILLA, *Aztec Thought and Culture*, 1963

'Listen, mother,' said Father Zossima. 'Once in olden times a holy saint saw in the Temple a mother like you weeping for her little one, her only one, whom God had taken. "Knowest thou not," said the saint to her, "how bold these little ones are before the throne of God? Verily, there are none bolder than they in the Kingdom of Heaven. 'Thou didst give us life, oh Lord,' they say, 'and scarcely had we looked upon it when Thou didst take it back again.' And so boldly they ask and ask again that God gives them at once the rank of angels . . ." That's what the saint said to the weeping mother of old. He was a great saint and he could not have spoken falsely.'

DOSTOEVSKY, *The Brothers Karamazov*

Neither the harps nor the crowns amused, nor the
    cherub's dove-winged races—
Holding hands forlornly the Children wandered beneath
    the Dome;
Plucking the radiant robes of the passers-by, and with
    pitiful faces
Begging what Princes and Powers refused:—'Ah,
    please will you let us go home?'

Over the jewelled floor, nigh weeping, ran to them
    Mary the Mother,
Kneeled and caressed and made promise with kisses,
    and drew them along to the gateway—
Yea, the all-iron unbribable Door which Peter must
    guard and none other—
Straightway She took the Keys from his keeping, and
    opened and freed them straightway.

Then to Her Son, Who had seen and smiled, She said:
'On the night that I bore Thee,
What didst Thou care for a love beyond mine or a
heaven that was not my arm?
Didst Thou push from the nipple, O Child, to hear
the angels adore Thee?
When we two lay in the breath of the kine?' And
He said:—'Thou hast done no harm.'

So through the Void the Children ran homeward
merrily hand in hand,
Looking neither to left nor right where the breathless
Heavens stood still;
And the Guards of the Void resheathed their swords,
for they heard the Command:
'Shall I that have suffered the children to come to me
hold them against their will?'

KIPLING, 'The Return of the Children'

When I am dead, and laid in grave,
And all my bones are rotten,
By this may I remembered be
When I should be forgotten.

On a girl's sampler, 1736

Animals

THE distinction is made at the outset that while, as we do, animals experience the physical pains attending death, they are spared (as far as one can tell) the pains of anticipation—of sensing in advance 'that last fated hateful journey to the vet', or the consequences of the hunter's gun, from which presumably the prey runs because it dislikes having things pointed at it.

Since we are ready to moralize about animals—can they really rise to suicide? Are they genuinely averse to incest?—it is only fair to allow them to moralize about us: in the voices of Cowper's Fop, Orlando Gibbons's sadly punning swan, La Fontaine's wounded bird and Johnson's virtuous vulture.

One might have expected more controversy over whether or not animals possess souls. But in the past souls were reserved to a favoured or right-minded minority even of the human race. And nowadays such is the suspicion attaching to the concept of 'soul' that animal lovers, while regarding their pets as full members of the family, may prefer not to pursue the question. Moreover there is a feeling in some quarters that animal lovers themselves lack whatever the contemporary replacement for souls is, being twisted, anti-social cranks quite prepared to have delinquents, the under-privileged, the disabled and so on chopped up to make pet food.

We are given these days to thinking of things as 'indivisible': freedom, for example. (If freedom of expression is to be guaranteed for the original thinker or creative artist, it must be made fully available to the professional pornographer, on the principle that all censorship is one censorship.) The theory strikes me as high-minded but quite unrealistic. We are lucky if we can come by good things in small portions, and they do not always fall into our lap as of 'right'. For this reason I hesitate to argue that kindness to living creatures of whatever kind is indivisible. And yet unkindness may well be: the man who kicks a cat is not unlikely to be found kicking his fellow men, especially if they are small and weak.

Animals are innocent in a sense that we are not, or not for very long, and certainly innocent of creating the conditions under which many of them live. It does not seem feasible that when God gave Adam dominion over the animals—the naming of them and the simultaneous understanding of their natures—he had in mind simply the harness and the stewpot. The conclusion to *The Ancient Mariner* is commonly judged a sad let-down, like a pinned-on badge promoting the activities of the RSPCA; and Coleridge himself thought the moral obtrusive for a work of pure imagination. But the imagination is never 'pure' for long at a stretch, so it might as well look to its morals. The proposition that 'he prayeth well, who loveth well both man and bird and beast' has at least stayed in our minds.

I take the liberty of closing this section with another and more admonitory moral, possibly a timely one, even though it assumes a sort of indivisibility in cause and effect.

~

The sense of death is most in apprehension,
And the poor beetle that we tread upon
In corporal sufferance finds a pang as great
As when a giant dies.

SHAKESPEARE, *Measure for Measure*

There were the sorrows of the dumb animals too—of the white angora, with a dark tail like an ermine's, and a face like a flower, who fell into a lingering sickness, and became quite delicately human in its valetudinarianism, and came to have a hundred different expressions of voice—how it grew worse and worse, till it began to feel the light too much for it, and at last, after one wild morning of pain, the little soul flickered away from the body, quite worn to death already, and now but feebly retaining it.

PATER, 'The Child in the House'

The animals that look at us like children
in innocence, in perfect innocence!
The innocence that looks at us! Like children
the animals, the simple animals,
have no idea why legs no longer work.

The food that is refused, the love of sleeping—
in innocence, in childhood innocence
there is a parallel of love. Of sleeping
they're never tired, the dying animals;
sick children too, whose play to them is work.

The animals are little children dying,
brash tigers, household pets—all innocence,
the flames that lit their eyes are also dying,
the animals, the simple animals,
die easily; but hard for us, like work!

GAVIN EWART (b. 1916), 'The Dying Animals

He came from Malta; and Eumelus says
He had no better dog in all his days.
We called him Bull; he went into the dark.
Along those roads we cannot hear him bark.

TYMNES (?2nd cent. BC), tr. Edmund Blunden

Your master dead, your life all out of tune,
You, Garsonek, on Thursday, died at noon.
You'd waited by his coffin for his call.
Beside his grave you'd watched the damp night fall.
You scented his familiar footsteps and
Followed until, by Charon's cruel command,
A small beast on no errand, you were tossed
Overboard whimpering; and the ferry crossed.
You dared not enter Lethe since you had feared
Water during your life: and so you reared
Back from the bank and barked and whined in vain,
And ran about and waited, and barked again.
Now you will howl forever on that shore.
Your own death holds you and the water's roar.
We, to reward the endless love you gave
Your master, give your corpse an earthen grave.

SAMUEL TWARDOWSKI (1600–60), 'Epitaph for a Dog', tr. Jerzy Peterkiewicz and Burns Singer

My dog lay dead five days without a grave
In the thick of summer, hid in a clump of pine
And a jungle of grass and honeysuckle-vine.
I who had loved him while he kept alive

Went only close enough to where he was
To sniff the heavy honeysuckle-smell
Twined with another odour heavier still
And hear the flies' intolerable buzz.

Well, I was ten and very much afraid.
In my kind world the dead were out of range
And I could not forgive the sad or strange
In beast or man. My father took the spade

And buried him. Last night I saw the grass
Slowly divide (it was the same scene
But now it glowed a fierce and mortal green)
And saw the dog emerging. I confess

I felt afraid again, but still he came
In the carnal sun, clothed in a hymn of flies,
And death was breeding in his lively eyes.
I started in to cry and call his name,

Asking forgiveness of his tongueless head.
. . . I dreamt the past was never past redeeming:
But whether this was false or honest dreaming
I beg death's pardon now. And mourn the dead.

RICHARD WILBUR (b. 1921), 'The Pardon'

On shallow straw, in shadeless glass,
Huddled by empty bowls, they sleep:
No dark, no dam, no earth, no grass—
Mam, get us one of them to keep.

Living toys are something novel,
But it soon wears off somehow.
Fetch the shoebox, fetch the shovel—
Mam, we're playing funerals now.

PHILIP LARKIN, 'Take One Home for the Kiddies'

'Ah, are you digging on my grave,
My loved one?—planting rue?'
—'No: yesterday he went to wed
One of the brightest wealth has bred.
"It cannot hurt her now," he said,
"That I should not be true." '

'Then who is digging on my grave?
    My nearest dearest kin?'
—'Ah, no: they sit and think, "What use!
What good will planting flowers produce?
No tendance of her mound can loose
    Her spirit from Death's gin." '

'But some one digs upon my grave?
    My enemy?—prodding sly?'
—'Nay: when she heard you had passed the Gate
That shuts on all flesh soon or late,
She thought you no more worth her hate,
    And cares not where you lie.'

'Then, who is digging on my grave?
    Say—since I have not guessed!'
—'O it is I, my mistress dear,
Your little dog, who still lives near,
And much I hope my movements here
    Have not disturbed your rest?'

'Ah, yes! *You* dig upon my grave . . .
    Why flashed it not on me
That one true heart was left behind!
What feeling do we ever find
To equal among human kind
    A dog's fidelity!'

'Mistress, I dug upon your grave
    To bury a bone, in case
I should be hungry near this spot
When passing on my daily trot.
I am sorry, but I quite forgot
    It was your resting-place.'

HARDY, 'Ah, Are You Digging on My Grave?'

Though once a puppy, and though Fop by name,
Here moulders one, whose bones some honour claim;
No sycophant, although of spaniel race!
And though no hound, a martyr to the chase!
Ye squirrels, rabbits, leverets, rejoice!
Your haunts no longer echo to his voice.

This record of his fate exulting view,
He died worn out with vain pursuit of you.
'Yes!' the indignant shade of Fop replies,
'And worn with vain pursuit, man also dies.'

WILLIAM COWPER (1731–1800), 'Epitaph on Fop'

I want him to have another living summer,
to lie in the sun and enjoy the *douceur de vivre*—
because the sun, like golden rum in a rummer,
is what makes an idle cat *un tout petit peu ivre*—

I want him to lie stretched out, contented,
revelling in the heat, his fur all dry and warm,
an Old Age Pensioner, retired, resented
by no one, and happinesses in a beelike swarm

to settle on him—postponed for another season
that last fated hateful journey to the vet
from which there is no return (and age the reason),
which must soon come—as I cannot forget.

GAVIN EWART, 'A 14-year-old convalescent cat in the winter'

My old cat is dead
Who would butt me with his head.
He had the sleekest fur,
He had the blackest purr.
Always gentle with us
Was this black puss,
But when I found him today
Stiff and cold where he lay,
His look was a lion's,
Full of rage, defiance:
O! he would not pretend
That what came was a friend
But met it in pure hate.
Well died, my old cat.

HAL SUMMERS (b. 1911), 'My Old Cat'

When we sat his mother on her tail, he mouthed her teat,
Slobbered a little, but after a minute
Lost aim and interest, his muzzle wandered,
He was managing a difficulty
Much more urgent and important. By evening
He could not stand. It was not
That he could not thrive, he was born
With everything but the will—
That can be deformed, just like a limb.
Death was more interesting to him.
Life could not get his attention.
So he died, with the yellow birth-mucus
Still in his cardigan.

TED HUGHES (b. 1930), from 'Sheep'

Only when she felt
The savage knife in her throat
Did the red veil
Explain the game
And she was sorry
She had torn herself
From the mud's embrace
And had hurried that evening
From the field so joyfully
Hurried to the yellow gate

VASKO POPA (b. 1922), 'Pig', tr. Anne Pennington

Though it will die soon
The voice of the cicada
Shows no sign of this.

BASHŌ (1644–94)

A rosy shield upon its back,
That not the hardest storm could crack,
From whose sharp edge projected out
Black pin-point eyes staring about;
Beneath, the well-knit cotte-armure
That gave to its weak belly power;
The clustered legs with plated joints
That ended in stiletto points;

The claws like mouths it held outside:
I cannot think this creature died
By storm or fish or sea-fowl harmed,
Walking the sea so heavily armed;
Or does it make for death to be
Oneself a living armoury?

ANDREW YOUNG (1885–1971), 'The Dead Crab'

Living in its shell
the conch never found it easy
to talk to man
now it is dead
it conveys the whole sea to the child's ear
full of wonder
full of wonder

RAYMOND QUENEAU (1903–76), 'Conch'

This, Lord, was an anxious brother and
a living diagram of fear: full of health himself,
he brought diseases like a gift
to give his hosts. Masked in a cat's moustache
but sounding like a bird, he was a ghost
of lesser noises and a kitchen pest
for whom some ladies stand on chairs. So,
Lord, accept our felt though minor guilt
for an ignoble foe and ancient sin:
the murder of a guest
who shared our board: just once he ate
too slowly, dying in our trap
from necessary hunger and a broken back.

ALAN DUGAN (b. 1923), from 'Funeral Oration for a Mouse'

When the mouse died, there was a sort of pity;
The tiny, delicate creature made for grief.
Yesterday, instead, the dead whale on the reef
Drew an excited multitude to the jetty.
How must a whale die to wring a tear?
Lugubrious death of a whale; the big
Feast for the gulls and sharks; the tug
Of the tide simulating life still there,
Until the air, polluted, swings this way
Like a door ajar from a slaughterhouse.

Pooh! pooh! spare us, give us the death of a mouse
By its tiny hole; not this in our lovely bay.
—Sorry we are, too, when a child dies:
But at the immolation of a race, who cries?

JOHN BLIGHT (b. 1913), 'Death of a Whale'

A winter seagull
In its life it has no home
In its death no grave

KATŌ SHŪSON (b. 1905)

HERE LIES TWEETER
WRAPPED IN SILK
THE LITTLE BIRD
DROWNED IN A
GLASS OF MILK

In a pets' cemetery, San Francisco, mentioned in Richard Brautigan, *The Tokyo–Montana Express*, 1980

A bird is tangled in a tree,
It flutters but it can't get free.
A black cat comes with stealthy tread,
Its claws are sharp, its eyes blood-red.
The poor bird sees it climbing higher,
Between the branches, ever nigher.
The bird reflects: What will be, will be,
And since the cat is bound to kill me,
I would not lose what time remains—
I'll warble unpremeditated strains
And trill awhile with wonted fervour.
The bird, it seems, does not lack humour!

WILHELM BUSCH (1832–1908), 'A bird is tangled in a tree'

The silver swan, who living had no note,
When death approached unlocked her silent throat;
Leaning her breast against the reedy shore,
Thus sung her first and last, and sung no more:
Farewell, all joys; O death, come close mine eyes;
More geese than swans now live, more fools than wise.

ORLANDO GIBBONS (1583–1625), 'The Silver Swan'

Time was when I was free as air,
The thistle's downy seed my fare,
My drink the morning dew;
I perch'd at will on ev'ry spray,
My form genteel, my plumage gay,
My strains for ever new.

But gaudy plumage, sprightly strain,
And form genteel, were all in vain,
And of a transient date;
For, caught and cag'd, and starv'd to death,
In dying sighs my little breath
Soon pass'd the wiry grate.

Thanks, gentle swain, for all my woes,
And thanks for this effectual close
And cure of ev'ry ill!
More cruelty could none express;
And I, if you had shown me less,
Had been your pris'ner still.

COWPER, 'On a Goldfinch Starved to Death in his Cage'

So the struck eagle, stretch'd upon the plain,
No more through rolling clouds to soar again,
View'd his own feather on the fatal dart,
And wing'd the shaft that quiver'd in his heart;
Keen were his pangs, but keener far to feel
He nursed the pinion which impell'd the steel;
While the same plumage that had warm'd his nest
Drank the last life-drop of his bleeding breast.

BYRON, *English Bards and Scotch Reviewers*

Mortally smitten by a feather'd dart,
A Bird bewailed his piteous destiny,
Sighing, as he saw what shaft had pierced his heart,
'Must we abet our own calamity?
Despiteful Humans! from our wings you tear
What speeds your deadly missiles to the aim;
Yet, ruthless breed! from mockery forbear:
Full oft your fortune is the same.
Of Adam's children, brother preying on brother,
One half will always arm the other.'

LA FONTAINE, 'The Wounded Bird', tr. Edward Marsh

A widow bird sate mourning for her love
  Upon a wintry bough;
The frozen wind crept on above,
  The freezing stream below.

There was no leaf upon the forest bare,
  No flower upon the ground,
And little motion in the air
  Except the mill-wheel's sound.

SHELLEY, *Charles the First*

But when men have killed their prey, said the pupil, why do they not eat it? When the wolf has killed a sheep, he suffers not the vulture to touch it till he has satisfied himself. Is not man another kind of wolf? Man, said the mother, is the only beast who kills that which he does not devour, and this quality makes him so much a benefactor to our species . . . When you see men in great numbers moving close together, like a flight of storks, you may conclude that they are hunting, and that you will soon revel in human blood. But still, said the young one, I would gladly know the reason of this mutual slaughter. I could never kill what I could not eat. My child, said the mother, this is a question which I cannot answer, though I am reckoned the most subtile bird of the mountain.

JOHNSON, *The Idler*, 22

Hi! handsome hunting man
Fire your little gun.
Bang! Now the animal
Is dead and dumb and done.
Nevermore to peep again, creep again, leap again,
Eat or sleep or drink again, Oh, what fun!

DE LA MARE, 'Hi!'

Sweet bird! though thou hast lost thy plumage, thou shalt fly to my mistress! Is it not better to be nibbled by her than mumbled by a cardinal? I, too, will feed on thy delicate beauty. Sweet bird! thy companion has fled to my mistress; and now thou shalt thrill the nerves of her master! Oh! doff, then, thy waistcoat of wine-leaves, pretty rover! and show me that bosom more delicious even than woman's. What gushes of rapture! What a flavour! How peculiar! Even how sacred! Heaven at once sends both manna and quails.

Another little wanderer! Pray follow my example! Allow me. All Paradise opens! Let me die eating ortolans to the sound of soft music!

BENJAMIN DISRAELI (1804–81), *The Young Duke*

The driver wasn't even aware
that he'd run over the little bird.
Suddenly it had a name
and address, a colour to its wings.
It lay in the middle of the street,
thrown onto its back,
feet lifted in a diagonal V.
Strange,
even truckdrivers noticed it now,
spread over it
a whistling tunnel. Finally
a pedestrian came
and gave it a last kick.

All this happened in broad daylight,
to the sound of buzzsaws
from a nearby carpenter's shop.
Meanwhile night has come.
I suppose the bird
is still there, clinging
to the gutter's edge.
I note it among the things
I should forget.

T. CARMI (b. 1925), 'Examination of Conscience Before Going to Sleep', tr. Stephen Mitchell

Two frogs I met in early childhood have lingered in my memory: I frightened one frog, and the other frog frightened me.

The frightened frog evinced fear by placing its two hands on its head: at least, I have since understood that a frog assumes this attitude when in danger, and my frog assumed it.

The alarming frog startled me, 'gave me quite a turn,' as people say, by jumping when I did not know it was near me . . .

But seeing that matters are as they are—because frogs and suchlike cannot in reason frighten us now—is it quite certain that no day will ever come when even the smallert, weakest, most grotesque, *wronged* creature will not in some fashion rise up in the Judgement with us to condemn us, and so frighten us effectually once for all?

CHRISTINA ROSSETTI, *Time Flies*

Mr Mould and his men had not exaggerated the grandeur of the arrangements. They were splendid. The four hearse-horses, especially, reared and pranced, and showed their highest action, as if they knew a man was dead, and triumphed in it. 'They break us, drive us, ride us; ill-treat, abuse, and maim us for their pleasure—But they die; Hurrah, they die!'

DICKENS, *Martin Chuzzlewit*

There were three ravens sat on a tree,
They were as black as they might be.

The one of them said to his make,
'Where shall we our breakfast take?'

'Down in yonder greenë field
There lies a knight slain under his shield.

'His hounds they lie down at his feet,
So well do they their master keep.

'His hawks they flie so eagerly,
There's no fowl dare come him nigh.

'Down there comes a fallow doe
As great with young as she might goe.

'She lifted up his bloudy head
And kist his wounds that were so red.

'She gat him up upon her back
And carried him to earthen lake.

'She buried him before the prime,
She was dead herself ere evensong time.

'God send every gentleman
Such hounds, such hawks, and such a leman.'

ANON., 'The Three Ravens'

When they saw Patroklos dead
—so brave and strong, so young—
the horses of Achilles began to weep;
their immortal natures were outraged
by this work of death they had to look at.
They reared their heads, tossed their manes,
beat the ground with their hooves,
and mourned Patroklos, seeing him lifeless, destroyed,
now mere flesh only, his spirit gone,
defenceless, without breath,
turned back from life to the great Nothingness.

Zeus saw the tears of those immortal horses and felt sorry.
'I shouldn't have acted so thoughtlessly
at the wedding of Peleus,' he said.
'Better if we hadn't given you as a gift,
my unhappy horses. What business did you have down there,
among pathetic human beings, the toys of fate?
You're free of death, you won't get old,
yet ephemeral disasters torment you.
Men have caught you in their misery.'
But it was for the eternal disaster of death
that those two gallant horses shed their tears.

CAVAFY, 'The Horses of Achilles', tr. Edmund Keeley and Philip Sherrard

Aristotle's horse having refused to mount his dam was induced to do so by the expedient of veiling her. On discovering his mistake he jumped intentionally from a hill and was killed by the fall, thus showing an antique moral discrimination unknown to his descendants.

Athenaeus in the *Deipnosophists*, a fascinating collection of anecdote, scandal, and gastronomic wisdom of the third century AD, retails another animal suicide springing from an even nicer and more easily outraged moral sense. 'A bird called the porphyrion [probably *Fulica porphyrio*, of the same family as the Common Coot], when it is kept in a house, watches those women who have husbands very closely; and has such instantaneous perception of anyone who commits adultery, that, when it perceives it, it gives notice of it to the master of the house, taking its own life by hanging itself.'

HENRY ROMILLY FEDDEN, *Suicide: A Social and Historical Study*

A doubt has been raised—whether brute animals ever commit suicide: to me it is obvious that they do not, and cannot. Some years ago, however, there was a case reported in all the newspapers of an old ram who committed suicide (as it was alleged) in the presence of many witnesses. Not having any pistols or razors, he ran for a short distance, in order to aid the impetus of his descent, and leapt over a precipice, at the foot of which he was dashed to pieces. His motive to the 'rash act', as the papers called it, was supposed to be mere *taedium vitae*.

THOMAS DE QUINCEY (1785–1859), 'Notes from the Pocket-Book of a late Opium-Eater'

If the irritated scorpion pierces itself with its sting (which is not at all certain), it is probably from an automatic, unreflecting reaction. The motive energy aroused by his irritation is discharged by chance and at random; the creature happens to become its victim, though it cannot be said to have had a preconception of the result of its action. On the other hand, if some dogs refuse to take food on losing their masters, it is because the sadness into which they are thrown has automatically caused lack of hunger; death has resulted, but without having been foreseen. Neither fasting in this case nor the wound in the other has been used as a means to a known effect. So the special characteristics of suicide as defined by us are lacking . . . A very small but highly suspicious number of cases may not be explicable in this way. For instance as reported by Aristotle, that of a horse who, realizing that he had been made to cover his dam without knowing the fact and after repeated refusals, flung himself intentionally from a cliff (*History of Animals*, IX, 47). Horse-breeders state that horses are by no means averse to incest.

DURKHEIM, *Suicide: A Study in Sociology*

They say, God wot!
She died upon the spot:
But then in spots she was so rich,—
I wonder which?

HOOD, 'On the Death of the Giraffe'

Small boy. 'Where do animals go when they die?'
Small girl. 'All good animals go to heaven, but the bad ones go to the Natural History Museum.'

Caption to drawing by E. H. Shepard, *Punch*, 1929

In paper case,
Hard by this place,
Dead a poor dormouse lies;
And soon or late,
Summoned by fate,
Each prince, each monarch dies.

Ye sons of verse,
While I rehearse,
Attend instructive rhyme;
No sins had Dor
To answer for,
Repent of yours in time.

ANON., 'Epitaph on a Dormouse, which some Children were to bury', 1765

LAST WORDS

Bird of Paradise: 'Home at last!'

Lemming: 'Thalassa! Thalassa!'

Mackerel: 'Mackerel Skies, receive me!'

Swan: 'Leda, where are you?'

Stallion: 'There will be mares' tails in heaven.'

Old Trout: 'Do stop tickling!'

Electric Eel: 'I must tell Faraday we got there first.'

Chameleon: 'How much darker I'm growing!'

Last Dodo: 'Now I'm extinct.'

Crocodile: 'No tears, *if* you please.'

Recorded by G. W. STONIER (b. 1903)

In a case where an animal is mortally wounded and cannot rise, the other members of the herd . . . circle it disconsolately several times, and if it is still motionless they come to an uncertain halt. They then face outward, their trunks hanging limply to the ground. After a while they may prod and circle again, and then again stand, facing outward. Eventually, if the fallen animal is dead, they move aside and just hang around . . . for several hours, or until nightfall, when they may tear out branches and grass clumps from the surrounding vegetation and drop these on and around the carcass, the younger elephants also taking part in this behaviour. They also scrape soil toward the carcass and then stand by, weaving restlessly from side to side. Eventually they move away from the area.

SYLVIA SIKES, *Natural History of the African Elephant*, 1970

I saw with open eyes
 Singing birds sweet
Sold in the shops
 For the people to eat,
Sold in the shops of
 Stupidity Street.

I saw in vision
 The worm in the wheat,
And in the shops nothing
 For people to eat;
Nothing for sale in
 Stupidity Street.

RALPH HODGSON (1871–1962), 'Stupidity Street'

Epitaphs, Requiems and Last Words

'THE history of common life seems as circumscribed as its moral attributes,' reflects a character in Peacock's *Melincourt* as he strolls in a churchyard, 'for the most extensive information I can collect from these gravestones is, that the parties married, lived in trouble, and died of a conflict between a disease and a physician.' The efforts of the tongue-tied can be touching, even so; as Tony Harrison admits, remembering his father:

> I've got the envelope that he'd been scrawling,
> misspelt, mawkish, stylistically appalling
> but I can't squeeze more love into their stone.

Epitaphs can be more informative and more engaging than those inspected by Mr Fax. With some irresistible exceptions, I have confined myself here to the class of epitaph, like the (ubiquitous) blacksmith's, the printer's and the one proposed by Brecht for himself, which issues out of the subjects' life-work. Whether or not they have been allowed a death of their own, they have been granted their own appropriate memorial. The same preference underlies the choice of Last Words: instances that look forwards as well as backwards and (as with Gainsborough, Adam Smith and Beethoven) predicate a continuity of occupation between this life and the next. 'I must work the works of him that sent me'—not only in the day that goes but in the night that is coming.

While there is a wealth of well-turned, just and moving epitaphs, some practitioners in the genre would seem to possess either no sense of humour—Captain Wedgwood's gamekeeper is apparently commended for shooting his master—or else an excess of it. Punning is rife on memorial tablets, possibly in a desperate attempt to 'personalize' the inscription, perhaps owing to a lack of anything substantial to record. Someone who lived to be a hundred is said to have died Young, that being his name; a person called Shallow is laid in a Deep grave; a Mrs Nott was Nott Alive and is Nott Dead; the poet Martial notes that the child Urbicus ('of the city') was 'Roman both by birth and name', as if a passport were needed for the other world; the seventeenth-century historian Thomas Fuller even punned on himself: 'Fuller's Earth'. In a more respectable epitaph for the affianced couple struck by lightning than the version printed earlier in this book, Pope describes them as 'Victims so pure Heav'n saw well pleas'd / And snatch'd them in Celestial fire', while in a relatively dignified example Crashaw commemorates a Dr Brooke thus:

> A *Brooke* whose stream so great, so good,
> Was lov'd, was honour'd as a flood:
> Whose Banks the Muses dwelt upon,
> More than their own Helicon:
> Here at length hath gladly found
> A quiet passage underground . . .

In *Encounter* (November 1961) Nigel Dennis cast grave doubts on the entire institution of Last Words. It does strain credulity that Rabelais should be credited with five completely dissimilar sets (of which only one is merely tetchy) and Heine with three, for it implies a wellnigh operatic loquacity. A doctor told Mr Dennis that he had attended five hundred death-beds without having heard a single memorable utterance, while a nurse explained that patients were sedated well in advance and consequently 'pass away without a word'. (Another nurse, however, admitted that patients often expressed gratitude for attentions received.) 'Just as the pneumatic tyre has driven the straw from the street of the dying,' Mr Dennis concluded, 'so has the hypodermic silenced the householder.'

It takes a brave, dedicated person to waive opiates for the sake of a final word or awareness, and no doubt in the past ghost writers have taken a hand and embellishments have crept in over the years. Yet we like to think that Marie-Antoinette, treading accidentally on the executioner's foot, did whisper 'Pardon, Monsieur' and that Andrew Bradford, an eighteenth-century Philadelphian newspaper publisher, did cry 'Oh Lord, forgive the errata!' And since actions occasionally speak louder than words, we hope that Petronius really did smash the more valuable pieces in his collection of pottery because he knew Nero coveted them.

Of the specimens printed here, I think it can be said that, if they are not literally authentic, they are authentically in character; and some of them, not exclusively pre-hypodermic, are known to be genuine. Among the latter is the query put to his son by Goronwy Rees: it constitutes the last words of this book.

Rich men, trust not in wealth,
Gold cannot buy you health,
Physic himself must fade.
All things, to end are made,
The plague full swift goes by,
I am sick, I must die:
Lord have mercy on us.

Beauty is but a flower,
Which wrinkles will devour,
Brightness falls from the air,
Queens have died young, and fair,
Dust hath closed Helen's eye.
I am sick, I must die:
Lord have mercy on us.

Strength stoops unto the grave,
Worms feed on Hector brave,
Swords may not fight with fate,
Earth still holds ope her gate.
Come, come, the bells do cry.
I am sick, I must die:
Lord have mercy on us . . .

THOMAS NASHE (1567–1601), *Summer's Last Will and Testament*

Hark, now everything is still,
The screech-owl, and the whistler shrill
Call upon our dame, aloud,
And bid her quickly don her shroud.
Much you had of land and rent,
Your length in clay's now competent.
A long war disturb'd your mind,
Here your perfect peace is sign'd.
Of what is't fools make such vain keeping?
Sin their conception, their birth weeping;
Their life a general mist of error,
Their death a hideous storm of terror.
Strew your hair with powders sweet,
Don clean linen, bathe your feet,
And (the foul fiend more to check)
A crucifix let bless your neck.
'Tis now full tide, 'tween night and day:
End your groan, and come away.

WEBSTER, *The Duchess of Malfi*

And here the precious dust is laid,
Whose purely temper'd clay was made
So fine, that it the guest betray'd.

Else the soul grew so fast within
It broke the outward shell of sin,
And so was hatch'd a cherubim.

In height it soar'd to God above;
In depth it did to knowledge move,
And spread in breadth to general love.

Before, a pious duty shin'd
To parents; courtesy behind;
On either side, an equal mind.

Good to the poor, to kindred dear,
To servants kind, to friendship clear:
To nothing but herself severe.

So, though a virgin, yet a bride
To every grace, she justified
A chaste polygamy, and died.

Learn from hence, Reader, what small trust
We owe this world, where virtue must,
Frail as our flesh, crumble to dust.

THOMAS CAREW (?1595–?1639), 'Maria Wentworth, at the age of 18'

Felix Randal the farrier, O he is dead then? my duty all ended,
Who have watched his mould of man, big-boned and hardy-handsome
Pining, pining, till time when reason rambled in it and some
Fatal four disorders, fleshed there, all contended?

Sickness broke him. Impatient he cursed at first, but mended
Being anointed and all; though a heavenlier heart began some
Months earlier, since I had our sweet reprieve and ransom
Tendered to him. Ah well, God rest him all road ever he offended!

This seeing the sick endears them to us, us too it endears.
My tongue had taught thee comfort, touch had quenched thy tears,
Thy tears that touched my heart, child, Felix, poor Felix Randal;

How far from then forethought of, all thy more boisterous years,
When thou at the random grim forge, powerful amidst peers,
Didst fettle for the great grey drayhorse his bright and battering sandal!

HOPKINS, 'Felix Randal'

Strew on her roses, roses,
  And never a spray of yew.
In quiet she reposes:
  Ah! would that I did too.

Her mirth the world required:
  She bathed it in smiles of glee.
But her heart was tired, tired,
  And now they let her be.

Her life was turning, turning,
  In mazes of heat and sound.
But for peace her soul was yearning,
  And now peace laps her round.

Her cabined, ample Spirit,
  It fluttered and failed for breath.
Tonight it doth inherit
  The vasty Hall of Death.

MATTHEW ARNOLD (1822–88), 'Requiescat'

Peace, peace! he is not dead, he doth not sleep—
He hath awakened from the dream of life—
'Tis we, who lost in stormy visions, keep
With phantoms an unprofitable strife,
And in mad trance, strike with our spirit's knife
Invulnerable nothings.—*We* decay
Like corpses in a charnel; fear and grief
Convulse us and consume us day by day,
And cold hopes swarm like worms within our living clay.

He has outsoared the shadow of our night;
Envy and calumny and hate and pain,
And that unrest which men miscall delight,
Can touch him not and torture not again;
From the contagion of the world's slow stain
He is secure, and now can never mourn
A heart grown cold, a head grown grey in vain;
Nor, when the spirit's self has ceased to burn,
With sparkless ashes load an unlamented urn . . .

SHELLEY, *Adonais: An Elegy on the Death of John Keats*

Fear no more the heat o' the sun,
 Nor the furious winter's rages;
Thou thy worldly task hast done,
 Home art gone and ta'en thy wages:
Golden lads and girls all must,
As chimney-sweepers, come to dust.

Fear no more the frown o' the great,
 Thou art past the tyrant's stroke;
Care no more to clothe and eat;
 To thee the reed is as the oak:
The sceptre, learning, physic, must
All follow this, and come to dust.

Fear no more the lightning-flash,
 Nor the all-dreaded thunder-stone;
Fear not slander, censure rash;
 Thou hast finish'd joy and moan:
All lovers young, all lovers must
Consign to thee, and come to dust.

No exorciser harm thee!
 Nor no witchcraft charm thee!
Ghost unlaid forbear thee!
 Nothing ill come near thee!
Quiet consummation have;
And renownèd be thy grave!

SHAKESPEARE, *Cymbeline*

I yesterday passed a whole afternoon in the churchyard, the cloisters, and the church, amusing myself with the tombstones and inscriptions that I met with in those several regions of the dead. Most of them recorded nothing else of the buried person, but that he was born upon one day, and died upon another: the whole history of his life being comprehended in those two circumstances, that are common to all mankind . . . Some of them were covered with such extravagant epitaphs, that, if it were possible for the dead person to be acquainted with them, he would blush at the praises which his friends have bestowed upon him. There are others so excessively modest, that they deliver the character of the person departed in Greek or Hebrew, and by that means are not understood once in a twelvemonth. In the poetical quarter, I found there were poets who had no monuments, and monuments which had no poets. I observed, indeed, that the present war had filled the church with many of these uninhabited monuments, which had been erected to the memory of persons whose bodies were perhaps buried in the plains of Blenheim, or in the bosom of the ocean.

JOSEPH ADDISON (1672–1719), 'Thoughts in Westminster Abbey'

While we rested ourselves on a horizontal monument, which was elevated just high enough to be a convenient seat, I observed that one of the gravestones lay very close to the church—so close that the droppings of the eaves would fall upon it. It seemed as if the inmate of that grave had desired to creep under the church-wall. On closer inspection, we found an almost illegible epitaph on the stone, and with difficulty made out this forlorn verse:

Poorly lived,
And poorly died,
Poorly buried,
And no one cried.

It would be hard to compress the story of a cold and luckless life, death, and burial into fewer words, or more impressive ones; at least, we found them impressive, perhaps because we had to re-create the inscription by scraping away the lichens from the faintly-traced letters. The grave was on the shady and damp side of the church, endwise towards it, the head-stone being within about three feet of the foundation-wall; so that, unless the poor man was a dwarf, he must have been doubled up to fit him into his final resting place. No wonder that his epitaph murmured against so poor a burial as this! His name, as well as I could make it out, was Treeo—John Treeo, I think—and he died in 1810, at the age of seventy-four. The gravestone is so overgrown with grass and weeds, so covered with unsightly lichens, and so crumbly with time and foul weather, that it is questionable whether anybody will ever be at the trouble of deciphering it again. But there is a quaint and sad kind of enjoyment in defeating (to such slight degree as my pen may do it) the probabilities of oblivion for poor John Treeo, and asking a little sympathy for him, half a century after his death.

NATHANIEL HAWTHORNE (1804–64), *Our Old Home*: Lillington churchyard, near Leamington Spa

Here lies father and mother and sister and I,
  We all died within the space of one short year;
They all be buried at Wimble, except I,
  And I be buried here.

Staffordshire churchyard

Lo, Hudled up, together Lye
Gray Age, Grene youth, White Infancy.
If Death doth Nature's Laws dispence,
And reconciles All Difference
Tis Fit, One Flesh, One House Should have
One Tombe, One Epitaph, One Grave:
And they that Liv'd and Lov'd Either,
Should Dye and Lye and Sleep together.

Good Reader, whether go or stay
Thou must not hence be Long Away

On William Bartholomew, d. 1662, his wife and several children, St John the Baptist's, Burford

Warm summer sun shine kindly here:
Warm summer wind blow softly here:
Green sod above lie light, lie light:
Good-night, Dear Heart: good-night, good-night.

Memorial to Clorinda Haywood, St Bartholomew's, Edgbaston

I sought my death, and found it in the womb;
I looked for life, and saw it was a shade;
I trod the earth, and knew it was my tomb;
And now I die, and now I am but made;
The glass is full, and now my glass is run;
And now I live, and now my life is done.

CHIDIOCK TICHBORNE (?1558–86), 'Written the Night before he was Beheaded'

Cowards fear to Die, but Courage stout,
Rather than Live in Snuff, will be put out.

RALEGH, 'On the Snuff of a Candle the night before he died'

Once you shone among the living as the Morning Star;
Among the dead you shine now, as the Evening Star.

PLATO

That Morn which saw me made a Bride,
The Ev'ning witnessed that I died.
Those holy lights, wherewith they guide
Unto the bed the bashful Bride,
Serv'd, but as Tapers, for to burn
And light my Relics to their urn.
This Epitaph, which here you see,
Supplied the Epithalamie.

HERRICK, 'Upon a Maid that Died the Day She Was Married', after Meleager (*fl.* 90 BC)

Someone is glad that I, Theodorus, am dead
Another will be glad when that someone is dead
We are all in arrears to death.

SIMONIDES, tr. Peter Jay

Remember Eubolus, who lived and died sober?
This is his grave. We might as well drink, then:
We'll all drop anchor in the same final harbour.

LEONIDAS OF TARENTUM, tr. Fleur Adcock

If fruits are fed on any beast
Let vine-roots suck this parish priest,
For while he lived, no summer sun
Went up but he'd a bottle done,
And in the starlight beer and stout
Kept his waistcoat bulging out.

Then Death that changes happy things
Damned his soul to water springs.

SYNGE, after 'Epitaph on François Rabelais' by Pierre de Ronsard (1524–85)

Her body dissected by fiendish men,
Her bones anatomized,
Her soul we trust has risen to God,
A place where few physicians rise.

On a victim of the 'resurrection men', ? early 19th century

Here lie I by the chancel door;
They put me here because I was poor.
The further in, the more you pay,
But here lie I as snug as they.

Devon tombstone

HEROES, and KINGS! your distance keep:
In peace let one poor Poet sleep,
Who never flatter'd Folks like you:
Let Horace blush, and Virgil too.

POPE, 'Epitaph. For One who would not be buried in Westminster Abbey'

While *Butler*, needy Wretch! was yet alive,
No gen'rous Patron would a Dinner give:
See him, when starv'd to Death and turn'd to Dust,
Presented with a Monumental Bust!
The Poet's Fate is here in Emblem shown;
He ask'd for Bread, and he receiv'd a Stone.

SAMUEL WESLEY (1691–1739), 'On the Setting Up of Mr Butler's Monument in Westminster Abbey'

Well, then, poor G—— lies under ground!
So there's an end of honest *Jack*.
So little Justice here he found,
'Tis ten to one he'll ne'er come back.

POPE, 'Epitaph on G——'; probably mock epitaph on John Gay

Life is a jest; and all things show it.
I thought so once; but now I know it.

JOHN GAY (1685–1732), 'My Own Epitaph'

Here Reynolds is laid, and to tell you my mind,
He has not left a better or wiser behind;
His pencil was striking, resistless and grand,
His manners were gentle, complying and bland;
Still born to improve us in every part,
His pencil our faces, his manners our heart:

To coxcombs averse, yet most civilly steering,
When they judged without skill he was still hard of hearing:
When they talk'd of their Raphaels, Correggios and stuff,
He shifted his trumpet, and only took snuff.

OLIVER GOLDSMITH (1728–74), *Retaliation*; on Sir Joshua Reynolds

Hard was thy fate in all the scenes of life,
As daughter, sister, parent, friend and wife,
But harder still in death thy fate we own,
Mourned by thy Godwin—with a heart of stone.

WILLIAM ROSCOE, on the publication in 1798 of William Godwin's *Memoirs* of Mary Wollstonecraft

François am I, heavy my lot,
Born in Paris not far from Pontoise;
A length of rope, a running knot,
Will teach my neck the weight of my arse.

VILLON, 'Quatrain'

Here lie Willie M——hie's banes,
O Satan, when ye tak him,
Gie him the schulin' o' your weans;
For clever Deils he'll mak 'em!

ROBERT BURNS (1759–96), 'On a Schoolmaster in Cleish Parish, Fifeshire'

I need no gravestone, but
If you need one for me
I would like it to bear these words:
He made suggestions. We
Carried them out.
Such an inscription would
Honour us all.

BRECHT, 'I need no gravestone', tr. Michael Hamburger

Interr'd beneath this Marble Stone,
Lie Saunt'ring JACK, and Idle JOAN.
While rolling Threescore Years and One
Did round this Globe their Courses run;

If Human Things went Ill or Well;
If changing Empires rose or fell;
The Morning past, the Evening came,
And found this Couple still the same . . .
  Nor Good, nor Bad, nor Fools, nor Wise;
They would not learn, nor could advise:
Without Love, Hatred, Joy, or Fear,
They led—a kind of—as it were:
Nor Wish'd, nor Car'd, nor Laugh'd, nor Cry'd:
And so They liv'd; and so They dy'd.

MATTHEW PRIOR (1664–1721), from 'An Epitaph'

Oh, fond attempt to give a deathless lot
To names ignoble, born to be forgot!
In vain, recorded in historic page,
They court the notice of a future age:
Those twinkling tiny lustres of the land
Drop one by one from Fame's neglecting hand:
Lethean gulfs receive them as they fall,
And dark oblivion soon absorbs them all.
So when a child, as playful children use,
Has burnt to tinder a stale last year's news,
The flame extinct, he views the roving fire—
There goes my lady, and there goes the squire.
There goes the parson, oh! illustrious spark,
And there, scarce less illustrious, goes the clerk!

COWPER, 'On Observing Some Names of Little Note Recorded in the "Biographia Britannica"'

Here lies Piron, a complete nullibiety,
Not even a Fellow of a Learned Society.

ALEXIS PIRON (1689–1773), 'My Epitaph'

This stone commemorates his name.
This grave received his tiny frame.
He's food for worms. To be precise,
One worm, one mouthful, would suffice.

IMMANUEL FRANCES (1618–*c.*1710), 'Epitaph on a Dwarf', tr. Hyam Maccoby

To the pious memory

of Ralph Quelche & Jane his wife

Who slept } together in
Now sleepe }

{ Bed by ye space of 40 years
{ Grave till Ct shall awaken them

He } fell asleep Ano Dni { 1629 }
Shee } { 16 }

beinge aged { 63 } yeares

For ye fruites of their { Labours }
{ Bodies }

they left { ye new Inn twice built
{ one only Son and two

{ at th^r owne charge
{ daughters

Their son being liberally bred in ye university of Oxon
Thought himself bound to erect this small monument

Of { their } piety towards { God }
{ his } { Them }

AN^o DNI 16

St Helen's church, Benson

All who come my grave to see
Avoid damp beds and think of me.

The grave of Lydia Eason, St Michael's, Stoke

Reader!
If thou hast a heart fam'd for
Tenderness and Pity, Contemplate
this Spot.
In which are deposited the Remains
of a Young Lady, whose artless Beauty,
innocence of Mind, and gentle Manners,
once obtained her the Love and
Esteem of all who knew her, But when
Nerves were too delicately spun to
bear the rude Shakes and Jostlings
which we meet in this transitory
World, Nature gave way; She sunk
and died a Martyr to Excessive
Sensibility.

On Sarah Fletcher, d. 1799, aged 29, Abbey of St Peter and St Paul, Dorchester

SACRED TO THE MEMORY OF
CAPTAIN ANTHONY WEDGWOOD
ACCIDENTALLY SHOT BY HIS GAMEKEEPER
WHILST OUT SHOOTING
"WELL DONE THOU GOOD AND FAITHFUL SERVANT"

'The epitaph of one obscure Wedgwood': Barbara and Hensleigh Wedgwood, *The Wedgwood Circle 1730–1897*, 1980

Erected to the Memory
of
John M^c^Farlane
Drown'd in the Water of Leith
By a few affectionate friends

Edinburgh

Here
lies inter'd
the mortal remains
of
JOHN HULM
Printer

who, like an old, worn-out type
battered by frequent use
reposes in the grave
But not without a hope
that at some future time
he might be cast in the mould of righteousness
And safely locked-up
in the chase of immortality

He was distributed from the board of life
on the 9th day of Sept. 1827
Aged 75

Regretted by his employers
and
respected by his fellow artists

St Michael's church, Coventry

My sledge and hammer lies
declin'd
My bellows too have lost
their wind:
My fire's extinct my coals
decay'd
And in the dust my vice
is laid;
My days are spent my glass
is run,
My nails are drove my work
is done

St Britius' church, Brize Norton

Here lies a shoemaker whose knife and hammer
Fell idle at the height of summer,
Who was not missed so much as when the rain
Of winter brought him back to mind again.

He was no preacher but his working text
Was *See all dry this winter and the next.*
Stand still. Remember his two hands, his laugh,
His craftsmanship. They are his epitaph.

JOHN ORMOND (b. 1923), 'At his Father's Grave'

Buffalo Bill's
defunct
who used to
ride a watersmooth-silver
stallion
and break onetwothreefourfive pigeonsjustlikethat
Jesus

he was a handsome man
and what i want to know is
how do you like your blueeyed boy
Mister Death

E. E. CUMMINGS (1894–1962), 'Buffalo Bill's

Malcolm Lowry
Late of the Bowery
His prose was flowery
And often glowery
He lived, nightly, and drank, daily,
And died playing the ukulele.

MALCOLM LOWRY (1909–57), 'Epitaph'

I shall soon be laid in the quiet grave—thank God for the quiet grave—O! I can feel the cold earth upon me—the daisies growing over me—O for this quiet—it will be my first.

KEATS, reported by Joseph Severn in a letter, 1821

Under the wide and starry sky
Dig the grave and let me lie:
Glad did I live and gladly die,
And I laid me down with a will.

This be the verse you grave for me:
Here he lies where he long'd to be;
Home is the sailor, home from sea,
And the hunter home from the hill.

ROBERT LOUIS STEVENSON (1850–94), 'Requiem'

If I have given you delight
By aught that I have done,
Let me lie quiet in that night
Which shall be yours anon:

And for the little, little, span
The dead are borne in mind,
Seek not to question other than
The books I leave behind.

KIPLING, 'The Appeal'

His dying was slow and harrowing. The final stroke had been preceded by one or two premonitory ones, each causing a diminution just marked enough for the still conscious intelligence to register it, and the sense of disintegration must have been tragically intensified to a man like James, who had so often and deeply pondered on it, so intently watched for its first symptoms. He is said to have told his old friend Lady Prothero, when she saw him after the first stroke, that in the very act of falling (he was dressing at the time) he heard in the room a voice which was distinctly, it seemed, not his own, saying: 'So here it is at last, the distinguished thing!' The phrase is too beautifully characteristic not to be recorded. He saw the distinguished thing coming, faced it, and received it with words worthy of all his dealings with life.

EDITH WHARTON, *A Backward Glance*, 1934 (on Henry James, d. 1916)

LAST WORDS

O, but they say the tongues of dying men
Enforce attention like deep harmony.
Where words are scarce they are seldom spent in vain,
For they breathe truth that breathe their words in pain.

SHAKESPEARE, *Richard II*

ARCHIMEDES (212 BC): (on being ordered by a Roman soldier to follow him) 'Wait till I have finished my problem.'

EMPEROR VESPASIAN (79): 'Alas! I suppose I am turning into a god.'

RABELAIS (1553): (a) 'I go to seek a great Perhaps'; (b) (after receiving extreme unction) 'I am greasing my boots for the last journey'; (c) 'Ring down the curtain, the farce is over!'; (d) (wrapping himself in his domino, or hooded cloak) '*Beati qui in domino moriuntur*': blessed are they that die in the Lord/in a domino.

BOILEAU (1711): 'It is a great consolation to a poet on the point of death that he has never written a line injurious to good morals.'

RAMEAU (1764): (to his confessor) 'What the devil are you trying to sing, monsieur le curé? Your voice is out of tune.'

ALBRECHT VON HALLER (1777): 'My friend, the artery ceases to beat.'

VOLTAIRE (1778): (as the bedside lamp flared up) 'What? The flames already?'

SAMUEL JOHNSON (1784): 'The next night he was at intervals delirious; and in one of those fits, seeing a friend at the bedside, he exclaimed, "What, will that fellow never have done talking poetry to me?" He recovered his senses before morning, but spoke little after this.' (Anon., *Life of Samuel Johnson, LL.D.*, 1786)

GAINSBOROUGH (1788): 'We are all going to Heaven, and Van Dyck is of the company.'

NICOLAS BEAUZÉE (grammarian, 1789): '*Mes chers amis, je m'en vais ou je m'en vas, car l'un et l'autre se dit ou se disent.*'

ADAM SMITH (1790): 'I believe we must adjourn this meeting to some other place.'

GOETHE'S MOTHER (1808): (to a servant girl bringing an invitation to a party) 'Say that Frau Goethe is unable to come, she is busy dying at the moment.'

ALEXANDER WILSON (ornithologist, 1813): 'Bury me where the birds will sing over my grave.'

BEETHOVEN (1827): 'I shall hear in Heaven.'

HEGEL (1831): 'Only one man ever understood me . . . And he didn't understand me.'

CUVIER (1832): (as leeches were being applied to him) 'Nurse, it was I who discovered that leeches have red blood.'

HEINE (1856): (a) 'God will pardon me. It's his métier'; (b) when asked by his doctor, '*Pouvez-vous siffler?*' [*siffler*: to whistle; to hiss], he replied, 'Alas no, not even a comedy of M. Scribe's!'; (c) 'Write . . . paper . . . pencil.'

PALMERSTON (1865): 'Die, my dear Doctor?—That's the last thing I shall do!'

COROT (1875): 'I hope with all my heart there will be painting in Heaven.'

DISRAELI (1881): (Queen Victoria having proposed to visit him) 'Why should I see her? She will only want me to give a message to Albert.'

JOYCE (1941): 'Does nobody understand?'

GERTRUDE STEIN (1946): 'What is the answer?' After a short silence she laughed and added, 'Then what is the question?'

GIDE (1951): 'I am afraid my sentences are becoming grammatically incorrect.'

RALPH VAUGHAN WILLIAMS (in a conversation some two weeks before his death in 1958, recorded by Sylvia Townsend Warner in a letter): 'If I were reincarnated, I added, I think I would like to be a landscape painter. What about you? Music, he said, music. But in the next world I shan't be doing music, with all the striving and disappointments. I shall be being it.'

JAMES THURBER (1961): 'God bless . . . God damn.'

GORONWY REES (1979): (to his son, Daniel) 'What shall I do next?'

ACKNOWLEDGEMENTS

THE editor and publishers gratefully acknowledge permission to reprint copyright material in this book as follows:

Aeschylus: 'Inexorable Death', trans. C. M. Bowra from Aeschylus' *Niobe*, from *The Oxford Book of Greek Verse in Translation*, edd. T. F. Higham and C. M. Bowra (1938). Reprinted by permission of Oxford University Press.

Anna Akhmatova: 'When a man dies . . .' from *Selected Poems*, trans. Richard McKane (Penguin European Poets, 1969), p. 74. Translation copyright © Richard McKane 1969. Reprinted by permission of Penguin Books Ltd.

Christopher Alexander: from *A Pattern Language: Towns, Buildings, Construction*. Copyright © 1977 by Christopher Alexander. Reprinted by permission of Oxford University Press Inc.

Michael Alexander: 'The funeral of Scyld Shefing' from *Beowulf*, trans. Michael Alexander (Penguin Classics, 1973). Copyright © Michael Alexander 1973. Reprinted by permission of Penguin Books Ltd.

A. Alvarez: from *The Savage God*. Reprinted by permission of George Weidenfeld & Nicolson Ltd.

Yehuda Amichai: 'My Father's Memorial Day' from *Amen*, trans. from the Hebrew by the author with Ted Hughes. Copyright © Yehuda Amichai 1977. Reprinted by permission of Oxford University Press and Harper & Row Inc.

Kingsley Amis: 'Delivery Guaranteed' from *Collected Poems*. Reprinted by permission of Hutchinson Publishing Group Ltd. and Jonathan Clowes Ltd.

Anon: 'I am a student nurse . . .' Copyright © 1970 American Journal of Nursing Co. By permission.

Sylvia Anthony: from *The Discovery of Death in Childhood and After* (Penguin Education, 1973). Copyright © Sylvia Anthony 1971. Reprinted by permission of Penguin Books Ltd. and Basic Books Inc.

Philippe Ariès: from *The Hour of Our Death*, trans. Helen Weaver (Allen Lane, 1981). This translation copyright © Alfred A. Knopf Inc. 1981. Reprinted by permission of Penguin Books Ltd. and Editions du Seuil, Paris.

Asclepiades: 'Why Hoard your Maidenhead?' from *Translations From The Greek Anthology* by Robert Allason Furness (1931). Reprinted by permission of the Estate of Robert A. Furness and Jonathan Cape Ltd.

Herbert Asquith: 'Nightfall'. Reprinted by permission of Sidgwick & Jackson Ltd.

W. H. Auden: 'Talking to Myself'. Copyright © 1972 by W. H. Auden, from *W. H. Auden: Collected Poems*, ed. Edward Mendelson. Reprinted by permission of Faber & Faber Ltd. and Random House Inc.

Charles Baudelaire: 'The Death of Lovers', trans. Roy Campbell (Harvill Press, 1952).

Samuel Beckett: from *Embers*. Copyright © 1959 by Grove Press Inc. Reprinted by permission of Faber & Faber Ltd. and Grove Press Inc.

Giuseppe Belli: 'The Last Judgement', trans. Anthony Burgess, from *Abba Abba*. Copyright © 1977 by Anthony Burgess. Reprinted by permission of Little, Brown & Company, Liana Burgess, and Faber & Faber Ltd.

Saul Bellow: from *Humboldt's Gift*. Reprinted by permission of Harriet Wasserman Literary Agency Inc.

John Betjeman: 'The Cottage Hospital' from *Collected Poems*. Reprinted by permission of John Murray (Publishers) Ltd.

John Blight: 'Death of a Whale' from *A Beachcomber's Diary*. Reprinted by permission of Angus & Robertson (UK) Ltd.

Edmund Blunden: 'The Midnight Skaters' from *Collected Poems* (The Bodley Head); 'He Came from Malta' by Tymnes, trans. Edmund Blunden, from *Halfway House* (Cobden-Sanderson Ltd). Reprinted by permission of A. D. Peters & Co. Ltd.

Jorge Luis Borges: 'On the Death of Francisco López Merino' from *In Praise of Darkness*, trans. Norman Thomas di Giovanni. Copyright © 1969, 1970, 1971, 1972, 1973, 1974 by Emece Editores S.A. and Norman Thomas di Giovanni. Reprinted by permission of E. P. Dutton Inc.

Tadeusz Borowski: excerpts from 'Auschwitz, Our Home: A Letter' from *This Way For The Gas, Ladies and Gentlemen*, trans. Barbara Vedder. First published in 1959 in *Wybór Opowiadań*, Państwowy Instytut Wydawniczy. Copyright © 1959 by Maria Borowski. This translation copyright © 1967 by Penguin Books Ltd. Reprinted by permission of Penguin Books Ltd. and Viking Penguin Inc.

Ronald Bottrall: from Dante's *Inferno*, trans. Ronald Bottrall (BBC Publications, 1966). By permission of the translator.

John Bowlby: from *Loss: Sadness and Depression*, Vol. iii of *Attachment and Loss* by John Bowlby. Copyright © 1980 by Tavistock Institute of Human Relations. Reprinted by permission of The Hogarth Press Ltd. and Basic Books Inc.

Richard Brautigan: 'Here Lies Tweeter (in a pets' cemetery, San Francisco)', from *The Tokyo–Montana Express*. Copyright © 1980 by Richard Brautigan. Reprinted by permission of Jonathan Cape Ltd. and Delacorte Press/Seymour Lawrence.

Bertolt Brecht: 'On the suicide of the refugee W.B.', trans. John Willett; 'I need no gravestone', trans. Michael Hamburger. Copyright © 1976 by Eyre Methuen Ltd. from *Bertolt Brecht: Poems 1913–1956*. Reprinted by permission of Methuen, London, and Methuen Inc., New York. By arrangement with Suhrkamp Verlag, Frankfurt 1. All rights reserved.

Alexander Sergeyevich Buturlin: quoted in *Tolstoy Remembered by his Son*, by Sergei Tolstoy, trans. Moura Budberg. Reprinted by permission of George Weidenfeld & Nicolson Ltd.

Albert Camus: from *The Myth of Sisyphus and Other Essays*, trans. Justin O'Brien. Copyright © 1955 by Alfred A. Knopf Inc.; from *The Plague*, trans. Stuart Gilbert. Copyright © 1948 by Stuart Gilbert. Reprinted by permission of Hamish Hamilton Ltd. and Alfred A. Knopf Inc.

T. Carmi: 'Examination of Conscience Before Going to Sleep' from *Somebody Like You* (1971), trans. Stephen Mitchell. Reprinted by permission of André Deutsch Ltd.

Dora Carrington: from *Carrington: Letters and Extracts from her Diaries*, ed. David Garnett. Reprinted by permission of the David Garnett Estate, the Sophie Partridge Trust, Jonathan Cape Ltd., and Holt, Rinehart & Winston Inc.

C. P. Cavafy: 'Lovely White Flowers'; 'The Horse of Achilles', from *Collected Poems*, trans. Edmund Keeley and Philip Sherrard, ed. George Savidis, trans. copyright © 1975 by Edmund Keeley and Philip Sherrard. Reprinted by permission of The Hogarth Press Ltd., the Author's Literary Estate, Deborah Rogers Ltd., and Princeton University Press.

Paul Celan: 'Ice, Eden' from *Poems*, trans. Michael Hamburger. Copyright © 1980 by Michael Hamburger. Reprinted by permission of Carcanet New Press Ltd. and Persea Books Inc.

Leslie Clarkson: from *Death, Disease and Famine in Pre-Industrial England*. Reprinted by permission of Gill & Macmillan, Dublin.

Hannah Closs: from *Tristan* (Dakers, 1940/Vanguard Press, N.Y., 1967/Popular Library, 1978).

Richard Cobb: from *Death in Paris, 1795–1801*. Copyright © Richard Cobb 1978. Reprinted by permission of Oxford University Press.

Confucius: from *The Analects of Confucius*, trans. Arthur Waley. Copyright 1938 by George Allen & Unwin Ltd. Reprinted by permission of George Allen & Unwin Ltd. and Macmillan Publishing Co. Inc.

Edward Conze: from *Buddhist Scriptures*, trans. Edward Conze (Penguin Classics, 1959). Copyright © Edward Conze 1959. Reprinted by permission of Penguin Books Ltd.

Jilly Cooper: from *Class*. Reprinted by permission of Methuen, London, Ltd.

Tristan Corbière: extract from 'The End', trans. C. F. MacIntyre, from *Les Amours Jaunes*. Copyright © 1954, 1982 by The Regents of the University of California. Reprinted by permission of The University of California Press.

F. M. Cornford: from *The Republic of Plato*, trans. F. M. Cornford (1941). Reprinted by permission of Oxford University Press.

Frances Cornford: 'Epitaph for a Reviewer'; 'All Souls' Night', from *Collected Poems*. Reprinted by permission of Hutchinson Publishing Group Ltd.

E. E. Cummings: 'Buffalo Bill's' from *Complete Poems 1913–62* and also in *Tulips and Chimneys*. Copyright 1923, 1925, and renewed 1951, 1953, by E. E. Cummings, copyright © 1973, 1976 by the Trustees for the E. E. Cummings Trust, copyright © 1973, 1976 by George James Firmage. Reprinted by permission of Granada Publishing and Liveright Publishing Corporation.

Dante: from *Paradiso*, trans. T. W. Ramsey (Hand & Flower Press, 1952).

N. J. Dawood: from *The Koran*, trans. N. J. Dawood (Penguin Classics, 4th r/e 1974). Copyright © N. J. Dawood 1956, 1959, 1966, 1968, 1974. Reprinted by permission of Penguin Books Ltd.

Paul Dehn: 'Ring-a-ring o' neutrons' from *Quake, Quake, Quake* from *The Fern on the Rock*. Copyright © Dehn Enterprises Ltd. 1965, 1976. Reprinted by permission of Hamish Hamilton Ltd.

Walter de la Mare: 'De Profundis'; 'Hi'. Reprinted by permission of The Literary Trustees of Walter de la Mare and The Society of Authors as their representative.

Peter de Vries: from *Comfort Me With Apples* (Little, Brown & Co.). Copyright 1952, 1953, © 1956 by Peter de Vries. Reprinted by permission of Laurence Pollinger Ltd.

Emily Dickinson: 'Because I could not stop for Death' (no. 712); 'I heard a fly buzz—when I died . . .' (no. 465); 'He scanned it—staggered— . . .' (no. 1062); 'Those—dying then, . . .' (no. 1551). Reprinted by permission of the publishers and the Trustees of Amherst College from *The Poems of Emily Dickinson*, ed. Thomas H. Johnson, Cambridge, Mass.: The Belknap Press of Harvard University Press. Copyright 1951, © 1955, 1979 by the President and Fellows of Harvard College.

Fyodor Dostoevsky; from *The Brothers Karamazov*, trans. from the Russian by Constance Garnett (1923). Reprinted by permission of Wm. Heinemann Ltd. and Macmillan Publishing Co. Inc.

Keith Douglas: 'Vergissmeinnicht' from *The Complete Poems of Keith Douglas*, ed. Desmond Graham. Copyright © Marie J. Douglas 1978. Reprinted by permission of Oxford University Press.

Alan Dugan: from 'Funeral Oration for a Mouse', copyright © 1961 by Alan Dugan, from *Collected Poems* (Yale University Press 1969/Faber 1970). Reprinted by permission of the author and Faber & Faber Ltd.

Emile Durkheim: from *Suicide: A Study in Sociology*, trans. John A. Spaulding and George Simpson. Copyright 1951 by The Free Press, a Corporation. Reprinted by permission of Routledge & Kegan Paul Ltd. and Macmillan Publishing Co. Inc.

Alice Thomas Ellis: 'To Joshua'. Reprinted by permission.

William Empson: 'Ignorance of Death' from *Collected Poems of William Empson.* Copyright 1949, 1977 by William Empson. Reprinted by permission of Chatto & Windus Ltd. and Harcourt Brace Jovanovich Inc.

Epicurus: from 'Letter to Menoeceus' from *Epicurus: The Extant Remains*, trans. Cyril Bailey (1926). Reprinted by permission of Oxford University Press.

Erasmus: from *Colloquies*, trans. Craig R. Thompson. Copyright © 1965 by The University of Chicago. Reprinted by permission of The University of Chicago Press.

W. Y. Evans-Wentz: 'Prayer for Guidance' from *The Tibetan Book of the Dead*, ed. W. Y. Evans-Wentz (3/e 1956). Reprinted by permission of Oxford University Press.

Gavin Ewart: 'The Dying Animals'; 'A 14-year-old convalescent cat in the winter', from *The New Ewart*. Reprinted by permission of Hutchinson Publishing Group Ltd.

Peter Farb: from *Humankind*. Copyright © 1978 by Peter Farb. Reprinted by permission of A. M. Heath & Co. Ltd. and Houghton Mifflin Co.

A. S. L. Farquharson: from *The Meditations of the Emperor Marcus Aurelius*, ed. and trans. A. S. L. Farquharson (1944). Reprinted by permission of Oxford University Press.

Henry Romilly Fedden: from *Suicide: A Social and Historical Study* (Wm. Heinemann). Reprinted by permission of David Higham Associates Ltd.

Raymond Firth: from *The Fate of the Soul* (1955). Reprinted by permission of Cambridge University Press.

Roy Fisher: 'As He Came Near Death' from *Poems 1955–1980*. Copyright © Roy Fisher 1980. Reprinted by permission of Oxford University Press.

Camille Flammarion: from *Death and Its Mystery*, trans. Latrobe Carroll (Fisher Unwin, 1922). Reprinted by permission of Ernest Benn.

E. M. Forster: 'What I Believe' from *Two Cheers For Democracy* (1951). Reprinted by permission of Edward Arnold (Publishers) Ltd.

Immanuel Frances: 'Epitaph on a Dwarf', trans. Hyam Maccoby, from *The Elek Book of Oriental Verse*. Reprinted by permission of Granada Publishing Ltd.

Michael Frayn: from *Constructions* (Wildwood House, 1974). Copyright © 1974 by Michael Frayn. Reprinted by permission of Elaine Greene Ltd.

Sigmund Freud: from 'Thoughts for the Times on War and Death', revised translation by James Strachey, from *The Standard Edition of the Complete Psychological Works of Sigmund Freud*, volume xiv (London: Hogarth Press Ltd.) and also in *The Collected Papers of Sigmund Freud*, vol. iv, ed. Ernest Jones, MD. Authorized translation, under the supervision of Joan Riviere, published in the United States by Basic Books Inc. by arrangement with The Hogarth Press Ltd. and The Institute of Psycho-Analysis, London. Reprinted by permission of the Sigmund Freud Copyrights Ltd., The Institute of Psycho-Analysis, The Hogarth Press Ltd., and Basic Books Inc.: from *Beyond the Pleasure Principle*, trans. and ed. James Strachey. Copyright © 1961 by James Strachey, W. W. Norton & Co. Inc., 1975. Reprinted by permission of the Sigmund Freud Copyrights Ltd., The Institute of Psycho-Analysis, The Hogarth Press, and W. W. Norton & Co. Inc.: from *The Interpretation of Dreams*, trans. and ed. James Strachey, published in the USA by Basic Books Inc. by arrangement with George Allen & Unwin Ltd. and The Hogarth Press Ltd. Reprinted by permission of George Allen & Unwin Ltd. and Basic Books Inc.

Robert Frost: from 'Home Burial' from *The Poetry of Robert Frost*, ed. Edward Connery Lathem. Copyright 1930, 1939, © 1969 by Holt, Rinehart, & Winston, copyright © 1958 by Robert Frost, copyright © 1967 by Lesley Frost Ballantine. Reprinted by permission of Jonathan Cape Ltd., the Estate of Robert Frost, and Holt, Rinehart, & Winston.

Christopher Fry: from *The Lady's Not for Burning* (1950). Reprinted by permission of Oxford University Press.
Roy Fuller: from 'Ghost Voice' from *The Reign of Sparrows*. Reprinted by permission of London Magazine Editions.
P. N. Furbank: from 'A Note on Death', *Proteus*, no. 2, Feb. 1978. Reprinted by permission of the author.
Théophile Gautier: 'Posthumous Coquetry', trans. Arthur Symons. Reprinted by permission of H. F. Read.
André Gide: from *The Coiners*, trans. Dorothy Bussy (1950). Reprinted by permission of Macmillan Publishing Co. Inc., for Cassell.
Oliver St John Gogarty: 'To Death' from *The Collected Poems of Oliver St John Gogarty*. Copyright © 1954 by the Devin Adair Co. Reprinted by permission of Oliver D. Gogarty and The Devin Adair Co., Connecticut.
Geoffrey Gorer: from *Death, Grief and Mourning in Contemporary Britain* (Hutchinson). Reprinted by permission of David Higham Associates Ltd.
Maxim Gorky: from *On Literature*, trans. Ivy Litvinov and J. Katzer (n/e 1974). Reprinted by permission of The Copyright Agency of the USSR.
Robert Graves: 'The Villagers and Death'; 'Penthesileia' from *Collected Poems* (Cassell). Reprinted by permission of the author and A. P. Watt Ltd.
Graham Greene: from *The Quiet American* (Heinemann 1961/Bodley Head 1973). Reprinted by permission of Laurence Pollinger Ltd. and Viking Penguin Inc.
Joyce Grenfell: from *Joyce—By Herself and Her Friends* (Macmillan/Futura). Reprinted by permission of Richard Scott Simon Ltd.
Geoffrey Grigson: 'A Sandy Burial' from *A Skull in Salop* (Macmillan, 1967). Reprinted by permission of the author.
Ivor Gurney: 'It Is Near Toussaints', and from 'The Silent One', from *Collected Poems of Ivor Gurney*, ed. P. J. Kavanagh. Copyright © Robin Haines, Sole Trustee of the Gurney Estate, 1982. Reprinted by permission of Oxford University Press.
J. C. Hall: 'Twelve Minutes' from *A House of Voices* (Chatto & Windus, 1973). Reprinted by permission of the author.
Manya Harari: from *Memoirs 1906–69* (Harvill Press, 1972).
Tony Harrison: 'Still', first published in *Times Literary Supplement*, 13 November 1981. Reprinted by permission of the author. 'I've got the envelope that he'd been scrawling, . . .' from *Continuous*. Reprinted by permission of Rex Collings Ltd.
Jaroslav Hašek: from *The Good Soldier Švejk*, trans. Cecil Parrott (Thomas Y. Crowell Co./Wm. Heinemann). Copyright © 1973 by Cecil Parrott. Reprinted by permission of Wm. Heinemann Ltd. and Harper & Row Inc.
Ronald Hayman: from *K: A Biography of Kafka*. Reprinted by permission of George Weidenfeld & Nicolson Ltd.
Anthony Hecht: 'Tarantula, or The Dance of Death' from *The Hard Hours*. Copyright © 1967 Anthony Hecht. Reprinted by permission of Oxford University Press and Atheneum Publishers.
Heinrich Heine: 'Anniversary' and 'It's going out', sections 11 and 18 of *Lazarus*; 'The heavenly fields of Paradise' section 11 of *Zum Lazarus*, collected in *The Lazarus Poems*, trans. Alistair Elliot and published by Mid Northumberland Arts Group in association with Carcanet Press. Reprinted by permission.
Zbigniew Herbert: 'Report from Paradise' from *Zbigniew Herbert: Selected Poems*, trans. Czesław Miłosz (Penguin Modern European Poets, 1968), p. 131. Translation copyright © Czesław Miłosz and Peter Dale Scott 1968. Reprinted by permission of Penguin Books Ltd.; 'What Mr Cogito Thinks about Hell' from *Zbigniew Herbert: Selected Poems*, trans. John Carpenter and Bogdana Carpenter. © John Carpenter and Bogdana Carpenter 1977. Reprinted by permission of Oxford University Press.

John Hersey: from *Hiroshima.* Copyright 1946 and renewed 1974 by John Hersey. Reprinted by permission of Alfred A. Knopf Inc. Originally appeared in the *New Yorker*.

Rosalind Heywood: from *Man's Concern With Death.* Reprinted by permission of Hodder & Stoughton Ltd.

John Hinton: from *Man's Concern With Death.* Reprinted by permission of Hodder & Stoughton Ltd.

Russell Hoban: from *Kleinzeit* (Cape). Reprinted by permission of David Higham Associates Ltd.

Ralph Hodgson: 'Stupidity Street' from *Collected Poems* (Macmillan, London) and in *Poems* (Macmillan, New York). Copyright 1917 by Macmillan Publishing Co. Inc., renewed 1945 by Ralph Hodgson. Reprinted by permission of Mrs Ralph Hodgson, Macmillan, London and Basingstoke, and Macmillan Publishing Co. Inc.

Hugo von Hofmannsthal: from *Death and the Fool* from *Poems and Verse Plays*, ed. Michael Hamburger. Bollingen Series XXXIII, vol. ii. Copyright © 1961 by Princeton University Press. Reprinted by permission of Routledge & Kegan Paul Ltd. and Princeton University Press.

Friedrich Hölderlin: 'To the Fates' from *Hölderlin: Poems and Fragments* (Cambridge University Press, 1980), trans. Michael Hamburger. Reprinted by permission of the translator and Cambridge University Press.

Miroslav Holub: 'Five Minutes After the Air Raid' from *Miroslav Holub: Selected Poems*, trans. Ian Milner and George Theiner (Penguin Modern European Poets, 1967), p. 40. Copyright © Miroslav Holub 1967. Translation copyright © Penguin Books 1967. Reprinted by permission of Penguin Books Ltd.

Horace: 'Hold! Pale Death . . .' from *Horace: Odes* (Hart-Davis), trans. James Michie. Reprinted by permission of the translator.

A. E. Housman: 'Is my team ploughing' from 'A Shropshire Lad'—Authorized edition—from *The Collected Poems of A. E. Housman.* Copyright 1939, 1940, © 1965 by Holt, Rinehart & Winston, copyright © 1967, 1968 by Robert E. Symons. Reprinted by permission of The Society of Authors as the literary representative of the Estate of A. E. Housman, Jonathan Cape Ltd., and of Holt, Rinehart & Winston Publ. Inc.

Ted Hughes: from 'Sheep' from *Season Songs* by Ted Hughes, illustrated by Leonard Baskin. Text copyright © 1975 by Ted Hughes. Reprinted by permission of Faber & Faber Ltd. and Viking Penguin Inc.

Alice James: from *The Diary of Alice James*, ed. Leon Edel (1965). Reprinted by permission of Dodd, Mead & Co. Inc. and the Wm. Morris Agency Inc.

Randall Jarrell: 'The Death of the Ball-Turret Gunner' from *Randall Jarrell: The Complete Poems.* Copyright © 1945, 1969 by Mrs Randall Jarrell, copyright renewed © 1973 by Mrs Randall Jarrell. Reprinted by permission of Faber & Faber Ltd. and Farrar, Straus & Giroux Inc.

Elizabeth Jennings: 'The Fear of Death' from *The Listener.* Reprinted by permission of David Higham Associates Ltd.

James Joyce: from *A Portrait of the Artist as a Young Man.* Copyright 1916 by B. W. Huebsch. Copyright renewed 1944 by Nora Joyce. Definitive text copyright © 1964 by the Estate of James Joyce. Reprinted by permission of Jonathan Cape Ltd., the Society of Authors as the literary representative of the Estate of James Joyce, and Viking Penguin Inc. From *Ulysses.* Copyright 1914, 1918 by Margaret Caroline Anderson and renewed 1942, 1946 by Nora Joseph Joyce. Reprinted by permission of The Bodley Head, Random House Inc., and The Society of Authors as the literary representative of the Estate of James Joyce.

C. G. Jung: 'The Soul and Death' from *The Collected Works of C. G. Jung*, trans. R. F. C. Hull, Bollingen Series XX, vol. viii: *The Structure and Dynamics of the*

Psyche. Copyright © 1960, 1969 by Princeton University Press. Reprinted by permission of Routledge & Kegan Paul Ltd. and Princeton University Press.

Donald M. Kaplan and Armand Schwerner: from *The Domesday Dictionary* (Simon & Schuster 1963). Reprinted by permission of Donald M. Kaplan.

Anne Karpf: from *The Observer*, 17 August 1980. Reprinted by permission.

R. Kastenbaum and R. Aisenberg: from *The Psychology of Death*. Reprinted by permission of Robert Kastenbaum.

Kazuko Yamada: 'The Wind' from *The Songs of Hiroshima*, trans. Miyao Ohara and D. J. Enright. By permission.

Søren Kierkegaard: from *The Sickness Unto Death*, trans. Howard V. Hong and Edna H. Hong © 1980 by Princeton University Press. By permission.

Maxine Hong Kingston: from *China Men*. Copyright © 1977, 1978, 1979, 1980 by Maxine Hong Kingston. Reprinted by permission of Pan Books Ltd. and Alfred A. Knopf Inc.

Rudyard Kipling: 'The Appeal', copyright 1939 by Caroline Kipling; 'The Refined Man' and 'Common Form' from 'Epitaphs of War', copyright 1919 by Rudyard Kipling; 'The Return of the Children'. All from *Rudyard Kipling's Verse: Definitive Edition*. Reprinted by permission of Doubleday & Company Inc. and of A. P. Watt Ltd., for the National Trust and Macmillan, London, Ltd.

Heinrich von Kleist: from *An Abyss Deep Enough: Letters of Kleist*, trans. Philip B. Miller. Reprinted by permission of E. P. Dutton Inc.

Jean de La Fontaine: 'Death and the Woodman'; 'The Wounded Bird' from *The Fables of Jean de La Fontaine*, trans. Edward Marsh. Reprinted by permission of Wm. Heinemann Ltd. and Harper & Row Inc.

Giuseppe di Lampedusa: from *The Leopard*, trans. A. Colquhoun. Copyright © 1960 by William Collins Sons & Co. Ltd. and Pantheon Books Inc. Reprinted by permission of Pantheon Books, a Division of Random House Inc., and Collins, Publishers.

Paul-Louis Landsberg: from *The Experience of Death and The Moral Problem of Suicide* (first published Rockliffe, 1953; Arno Press repr., 1977), trans. Cynthia Rowland.

Lao Tzu: from *Tao te Ching*, trans. Witter Bynner as *The Way of Life According to Lao Tzu*. Reprinted by permission of The Lyrebird Press Ltd.

Philip Larkin: 'Take One Home for the Kiddies' from *The Whitsun Weddings*. Reprinted by permission of Faber & Faber Ltd. 'Heads in the Women's Ward' and extract from 'Aubade'. Reprinted by permission of the author.

Christopher Leach: from *Letter to a Younger Son*. Copyright © 1981 by Christopher Leach. Reprinted by permission of J. M. Dent & Sons Ltd. and Harcourt Brace Jovanovich Inc.

Laurie Lee: from *Cider with Rosie*. Reprinted by permission of The Hogarth Press Ltd.

Leonidas of Tarentum: trans. Fleur Adcock, from *The Greek Anthology*, ed. Peter Jay (Penguin Classics, r/e 1981). Copyright © Peter Jay, 1973, 1981. Reprinted by permission of Penguin Books Ltd.

Miguel Léon-Portilla: from *Aztec Thought and Culture: A Study of the Ancient Nahuatl Mind*, trans. Jack Emory Davis. Copyright © 1963 by the University of Oklahoma Press. Reprinted by permission.

G. E. Lessing: from *Hamburgische Dramaturgie*, trans. Idris Parry, *PN Review*, 23 (1981). Reprinted by permission of Carcanet Press Ltd.

Claude Lévi-Strauss: from *Tristes Tropiques*, trans. John and Doreen Weightman (copyright © 1955 by Librairie Plon). Trans. copyright © 1973 by Jonathan Cape Ltd. Reprinted by permission of Jonathan Cape Ltd. and Atheneum Publishers.

C. S. Lewis: from *A Grief Observed*. Copyright © 1961 by N. W. Clerk. Reprinted by permission of Faber & Faber Ltd. and The Seabury Press Inc.

Robert Jay Lifton: from *Six Lives/Six Deaths*. Copyright © 1979 by Robert Jay Lifton, Shūichi Katō and Michael R. Reich. Reprinted by permission of Yale University Press.

Michael Longley: 'Wounds' from *An Exploded View*. Reprinted by permission of Victor Gollancz Ltd.

Federico García Lorca: 'Farewell' trans. W. S. Merwin, from *Selected Poems of Federico García Lorca*. Copyright © 1955 by New Directions Publishing Corp. Reprinted by permission of New Directions Publishing Corp.

Robert Lowell: 'Villon's Prayer for his Mother' from *Imitations* by Robert Lowell. Copyright © 1958, 1959, 1960, 1961 by Robert Lowell. Reprinted by permission of Faber & Faber Ltd. and Farrar, Straus & Giroux Inc.

Malcolm Lowry: 'Epitaph' from *Selected Poems*. Copyright © 1962 by Margerie Lowry. Reprinted by permission of City Lights Books.

Lucilius: 'Eutychides is dead, . . .', trans. Humbert Wolfe, from *Others Abide* (Ernest Benn, 1927). Reprinted by permission of Miss Ann Wolfe.

Norman MacCaig: 'Every Day'. Reprinted by permission of the author.

Robert W. MacKenna: from *The Adventure of Death*. Reprinted by permission of John Murray (Publishers) Ltd.

John McManners: from *Death and the Enlightenment*. © John McManners 1981. Reprinted by permission of Oxford University Press.

Louis MacNeice: 'Tam Cari Capitis' from *The Collected Poems of Louis MacNeice*. Reprinted by permission of Faber & Faber Ltd.

Maurice Maeterlinck: from *Death*, trans. Alexander Teixeira de Mattos. Reprinted by permission of Methuen, London, Ltd.

Bronisław Malinowski: from *Magic, Science and Religion*. Reprinted by permission of Souvenir Press Ltd.

Nadezhda Mandelstam: from *Hope Against Hope*, trans. Max Hayward (Harvill Press, 1970).

Osip Mandelstam: 'What Street is This' from *Osip Mandelstam: Selected Poems*, trans. Clarence Brown and W. S. Merwin. © Clarence Brown and W. S. Merwin 1973. Reprinted by permission of Oxford University Press & Atheneum Publishers.

Thomas Mann: from *Doctor Faustus*, trans. H. T. Lowe-Porter. Copyright 1948 by Alfred A. Knopf Inc.; from *The Magic Mountain*, trans. H. T. Lowe-Porter, copyright 1927 and renewed 1955 by Alfred A. Knopf Inc. Copyright 1952 by Thomas Mann; from *A Sketch of My Life*, trans. H. T. Lowe-Porter. Published 1960 by Alfred A. Knopf Inc. All rights reserved. Reprinted by permission of Martin Secker & Warburg Ltd. and Alfred A. Knopf Inc.

Katherine Mansfield: from 'Prelude'. Copyright 1920 by Alfred A. Knopf Inc. and renewed 1948 by John Middleton Murry, from *The Short Stories of Katherine Mansfield*. Reprinted by permission of Alfred A. Knopf Inc.

Jacques Maritain: from *Man's Destiny in Eternity*, ed. Arthur H. Compton. Copyright © 1949 by F. Lyman Windolph and Farmer's Bank and Trust Co. of Lancaster, trustees under the will of M. T. Garvin. Reprinted by permission of Beacon Press Inc.

Don Marquis: from 'ghosts' from *archy and mehitabel*. Copyright 1927 by Doubleday & Co. Inc. Reprinted by permission of Faber & Faber Ltd. and Doubleday & Co. Inc.

Martial: 'For Erotion's Grave', trans. F. A. Wright. Reprinted by permission of Routledge & Kegan Paul Ltd.

Edgar Lee Masters: 'Franklin Jones'; 'Chase Henry'; and 'Ollie McGee' from *Spoon River Anthology* (Macmillan Publishing Co. Inc.). Reprinted by permission of Eileen C. Masters.

Somerset Maugham: as quoted in *Escape from the Shadows* by Robin Maugham. Reprinted by permission of Hodder & Stoughton Ltd.

Micheline Maurel: from *Ravensbrück*, trans. Margaret S. Summers (Blond & Briggs, London, 1958). Reprinted by permission of Frederick Muller Ltd.

Derwent May: 'A Child in the 80s'. Reprinted by permission of the author.

Edna St Vincent Millay: 'Passer Mortuus Est' from *Collected Poems* (Harper & Row). Copyright 1921, 1948 by Edna St Vincent Millay. Reprinted by permission of Norma Millay Ellis, Literary Executor.

Czesław Miłosz: 'On the Other Side' from *Selected Poems*, by Czesław Miłosz, trans. Jan Darowski. Copyright © 1973 by the author. Reprinted by permission of the Continuum Publishing Company. 'The Fall', copyright © 1978 by Czesław Miłosz, trans. the author and Lillian Vallee, from *Bells in Winter*. Reprinted by permission of the Ecco Press.

Jessica Mitford: from *The American Way of Death*, copyright © 1963, 1978 by Jessica Mitford. Reprinted by permission of Penguin Books Ltd. and Simon & Schuster, a Division of Gulf & Western Corp.

Eugenio Montale: from 'Xenia I' from *New Poems*, trans. G. Singh. Copyright © 1970, 1972 by Eugenio Montale and G. Singh. Reprinted by permission of Chatto & Windus Ltd. and New Directions Publ. Corp. 'The Inhuman' from *It Depends: A Poet's Notebook*, trans. G. Singh. Copyright © 1977 by Arnoldo Mondadori Editore S.P.A., © 1980 by G. Singh. Reprinted by permission of New Directions Publ. Corp.

Christian Morgenstern: 'The Rabbi' from *Gallows Song*, trans. W. D. Snodgrass and Lore Segal. Reprinted by permission of the translators.

John Morley: from *Death, Heaven and the Victorians* (1971). Reprinted by permission of Macmillan Publishing Co. Inc., for Cassell.

Robert Musil: from *The Man Without Qualities*, trans. Eithne Wilkins and Ernst Kaiser (Secker). Reprinted by permission of Martin Secker & Warburg Ltd. and Alfred A. Knopf Inc.

Vladimir Nabokov: from *Pale Fire*. Reprinted by permission of George Weidenfeld & Nicolson Ltd.

Nakajini Tadasha: from *The Divine Wind*, trans. Roger Pineau. Reprinted by permission of Roger Pineau.

R. K. Narayan: from *My Days*. Reprinted by permission of the author, Chatto & Windus Ltd. and Anthony Sheil Associates.

St Elmo Nauman, Jr: from *Dictionary of Asian Philosophies*. Reprinted by permission of Routledge & Kegan Paul Ltd. and Philosophical Library Inc.

Flann O'Brien: from *At Swim-Two-Birds*. Reprinted by permission of Granada Publishing Ltd. and A. M. Heath & Co. Ltd., for the Estate of the late Flann O'Brien.

John Ormond: 'At His Father's Grave' from *Requiem and Celebration*. Reprinted by permission of Christopher Davies Ltd.

George Orwell: 'As I Please' from *The Collected Essays, Journalism and Letters of George Orwell*, vol. iii. Copyright © 1968 by Sonia Brownell Orwell. Reprinted by permission of Harcourt Brace Jovanovich Inc. and of A. M. Heath & Co. Ltd., for the Estate of the late Sonia Brownell Orwell, and Martin Secker & Warburg Ltd.

Harold Owen: from *Journey to Obscurity*, vol. iii © Oxford University Press 1965. Reprinted by permission of Oxford University Press.

Dan Pagis: 'Written in pencil in the sealed railway-car' from *Selected Poems*, trans. Stephen Mitchell. Reprinted by permission of Carcanet Press Ltd.

Dorothy Parker: 'Partial Comfort' from *The Collected Dorothy Parker*. Reprinted by permission of Gerald Duckworth & Co. Ltd.

Boris Pasternak: from *Doctor Zhivago*, trans. Max Hayward and Manya Harari. Copyright © 1958 by William Collins Sons & Co. Ltd. and Pantheon Books Inc.

Reprinted by permission of Pantheon Books, a Division of Random House Inc., and Harvill Press Ltd.

Cesare Pavese: from *This Business of Living: A Diary, 1935–1950*, ed. and trans. A. E. Murch. Reprinted by permission of Peter Owen Ltd., London.

Petrarch: 'Triumphs' from *Petrarch and His World*, trans. Morris Bishop. Reprinted by permission of Chatto & Windus Ltd. and Indiana University Press.

Lily Pincus: from *Death and the Family: The Importance of Mourning*. Copyright © 1974 by Lily Pincus. Reprinted by permission of Faber & Faber Ltd. and Pantheon Books Inc., a Division of Random House Inc.

Sylvia Plath: from 'Berck-Plage' from *Ariel*, © 1965 by Ted Hughes. Reprinted by permission of Olwyn Hughes.

Po Chü-i: 'Last Poem' from *Chinese Poems*, trans. Arthur Waley (1946). Reprinted by permission of George Allen & Unwin Ltd.

Peter Porter: 'Evolution' from *Preaching to the Converted*, © Oxford University Press 1972. Reprinted by permission of Oxford University Press.

Brian Power: speaking to his amah, in Tientsin, *c*.1925. By permission.

T. F. Powys: from 'A Dumb Animal' from *Rosie Plum and Other Stories*. Reprinted by permission of the Author's Literary Estate and Chatto & Windus Ltd.

Jacques Prévert: 'Familial', trans. D. J. Enright, from *Paroles*. © Editions Gallimard 1949. By permission of Editions Gallimard, Paris.

J. B. Priestley: from *Over the Long High Wall*. Reprinted by permission of Wm. Heinemann Ltd.

Marcel Proust: from *Remembrance of Things Past*, trans. C. K. Scott Moncrieff and Terence Kilmartin. Translation copyright © 1981 Random House Inc. and Chatto & Windus Ltd. Reprinted by permission of Chatto & Windus Ltd. and Random House Inc.

Raymond Queneau: 'Conch' ('Buccin'), trans. D. J. Enright. By permission of Editions Gallimard, Paris.

John Crowe Ransom: 'Dead Boy', 'Janet Waking'. Copyright 1927 by Alfred A. Knopf Inc., renewed 1955 by John Crowe Ransom, from *Selected Poems*. Reprinted by permission of Laurence Pollinger Ltd. and Alfred A. Knopf Inc.

Maurice Richardson and Philip Toynbee: from *Thanatos: A Modern Symposium*. Reprinted by permission of Victor Gollancz Ltd.

Rainer Maria Rilke: from *The Notebook of Malte Laurids Brigge*, trans. John Linton; from *The Book of Hours*, trans. J. B. Leishman; 'The Death of the Beloved' from *Rilke: New Poems*, trans. J. B. Leishman. Reprinted by permission of The Hogarth Press Ltd. and the Author's Literary Estate.

Joachim Ringelnatz: 'Ambition', trans. Christopher Middleton. Reprinted by permission of the translator.

Michael Roberts: ' "Already" said my Host' from *Collected Poems*. Reprinted by permission of Faber & Faber Ltd.

Gabriel Ronay: from *The Dracula Myth*. © Gabriel Ronay 1972. Reprinted by permission of the author and A. P. Watt Ltd.

Tadeusz Różewicz: 'Massacre of the Innocents' from *Faces of Anxiety* (1969), trans. Jan Darowski. Reprinted by permission of André Deutsch Ltd.

Jalāl al-Dīn Rūmī: poem no. 118 from *The Mystical Poems of Rūmˉ*, trans. A. J. Arberry. Copyright © 1968 by A. J. Arberry. Reprinted by permission of The University of Chicago Press.

Bertrand Russell: from *Unpopular Essays* (1950). Copyright 1950 by Bertrand Russell. Reprinted by permission of George Allen & Unwin Ltd. and Simon & Schuster Inc.

N. K. Sandars: from *The Epic of Gilgamesh*, trans. N. K. Sandars (Penguin Classics, r/e 1972). Copyright © N. K. Sandars 1960, 1964, 1972. Reprinted by permission of Penguin Books Ltd.

George Santayana: from *The Life of Reason*, abridged by Daniel M. Cory. Copyright 1953 Daniel M. Cory, copyright renewed 1981 Margot Cory. Reprinted by permission of Constable Publishers and Charles Scribner's Sons.

Jean-Paul Sartre: from *Words*, trans. Irene Clephane. Reprinted by permission of George Braziller Inc.

Siegfried Sassoon: from *Memoirs of an Infantry Officer*. Reprinted by permission of Faber & Faber Ltd., The K. S. Giniger Co. Inc. and Stackpole Books.

Vernon Scannell: 'Felo de Se' from *New and Collected Poems*. Reprinted by permission of Robson Books Ltd.

Seneca: from *Ad Marciam de Consolatione*, trans. J. W. Basore. Reprinted by permission of The Loeb Classical Library (Harvard University Press: Wm. Heinemann Ltd.).

Bernard Shaw: from *Man and Superman*. Reprinted by permission of The Society of Authors on behalf of the Bernard Shaw Estate.

Shen Ts'ung-wen: 'On an uprising in Fenghuang in 1911', trans. Nieh Hua-ling, from *Shen Ts'ung-wen*. Copyright 1972 by Twayne Publishers Inc. Reprinted by permission of Twayne Publishers, a division of G. K. Hall & Co., Boston.

Sylvia Sikes: from *Natural History of the African Elephant*. Reprinted by permission of George Weidenfeld & Nicolson Ltd.

Simonides: trans. Peter Jay, from *The Greek Anthology*, ed. Peter Jay (Penguin Classics, r/e 1981). Copyright © Peter Jay 1973, 1981. Reprinted by permission of Penguin Books Ltd.

Lucy Jane Simpson: 'Sheep's Skull', *Daily Mirror* Children's Literary Competition, 1971. By permission.

Robin Skelton: 'This is Anacreon's grave . . .', trans. Robin Skelton, in *200 Poems from the Greek Anthology* (Methuen). Reprinted by permission of the translator.

Stevie Smith: from *Novel on Yellow Paper* (n/e Virago, 1980). Reprinted by permission of James MacGibbon, Literary Executor.

Alexander Solzhenitsyn: '*We* Will Never Die', trans. Michael Glenny from *Stories and Prose Poems* (1971). Reprinted by permission of The Bodley Head Ltd.

Bernard Spencer: 'Sarcophagi' from *Bernard Spencer: Collected Poems*, ed. Roger Bowen. © Mrs Anne Humphreys 1981. Reprinted by permission of Oxford University Press.

H. J. Spinden: 'Tewa song', trans. H. J. Spinden, from *Songs of the Tewa* (1933). Reprinted by permission of AMS Press Inc.

Stendhal: from *Love*, trans. G. and K. Sale. Reprinted by permission of The Merlin Press Ltd.

Wallace Stevens: 'The Worm at Heaven's Gate'. Copyright 1923 and renewed 1951 by Wallace Stevens, from *The Collected Poems of Wallace Stevens*. Reprinted by permission of Faber & Faber Ltd. and Alfred A. Knopf Inc.

G. W. Stonier: 'Last Words'. By permission.

Tom Stoppard: from *Rosencrantz and Guildenstern are Dead*. Copyright © 1967 by Tom Stoppard. Reprinted by permission of Faber & Faber Ltd. and Grove Press Inc.

Hal Summers: 'My Old Cat' from *Tomorrow is My Love*. © Hal Summers 1978. Reprinted by permission of Oxford University Press.

Jules Supervielle: 'Whisper in agony' (from *Le Forçat innocent*), copyright Editions Gallimard 1930. Trans. D. J. Enright. By permission of Editions Gallimard.

David Sutton: 'Not to be Born' from *Absences and Celebrations*. Reprinted by permission of Chatto & Windus Ltd.

Rabindranath Tagore: from *Gitanjali*. Reprinted by permission of Macmillan, London and Basingstoke.

Cahit Sitki Taranci: 'One of the Dead Speaks', trans. Nermin Menemencioğlu, from *The Penguin Book of Turkish Verse* (1978). Reprinted by permission of the translator.

Allen Tate: from 'Ode to the Confederate Dead' from *Collected Poems 1919–1976*. Copyright © 1952, 1953, 1970, 1977 by Allen Tate. Copyright 1931, 1932, 1937, 1948 by Charles Scribner's Sons. Copyright renewed © 1959, 1960, 1965 by Allen Tate. Reprinted by permission of Farrar, Straus & Giroux Inc. and Faber & Faber Ltd.

Dylan Thomas: 'Do not go gentle into that good night', from *Collected Poems* (J. M. Dent). Reprinted by permission of David Higham Associates Ltd.

Leo Tolstoy: from *War and Peace*, trans. Louise and Aylmer Maude (1933). Reprinted by permission of Oxford University Press. From *Anna Karenina*, trans. Rosemary Edmonds (Penguin Classics, r/e 1978). Copyright 1954, 1978 by Rosemary Edmonds. Reprinted by permission of Penguin Books Ltd.

Arnold Toynbee: from *Man's Concern With Death*. Reprinted by permission of Hodder & Stoughton Ltd.

Marina Tsvetayeva: 'I know the truth' from *Marina Tsvetayeva: Selected Poems*, trans. Elaine Feinstein in association with Angela Livingstone (2/e 1981). © Oxford University Press 1971 and Elaine Feinstein 1981. Reprinted by permission of Oxford University Press.

Samuel Twardowski: 'Epitaph for a Dog' from *Five Centuries of Polish Poetry 1450–1970*, trans. Jerzy Peterkiewicz and Burns Singer (2/e 1970). © Jerzy Peterkiewicz and Burns Singer 1960, 1970. Reprinted by permission of Oxford University Press.

Raage Ugaas: 'A lament for his wife', trans. B. W. Andrzejewski and I. M. Lewis, from *Somali Poetry: An Introduction* (Oxford University Press, 1964). Reprinted by permission of the translators.

Miguel de Unamuno: from *Selected Works of Miguel de Unamuno*, Bollingen Series LXXXV, vol. IV: *The Tragic Sense of Life in Men and Nations*, trans. Anthony Kerrigan. Copyright © 1972 by Princeton University Press. Reprinted by permission of Routledge & Kegan Paul Ltd. and Princeton University Press.

Giuseppe Ungaretti: 'Watch' from *Selected Poems*, trans. Patrick Creagh (Penguin Modern European Poets, 1971), p. 28. Translation copyright © Patrick Creagh 1971. Reprinted by permission of Penguin Books Ltd.

Paul Valéry: from 'The Graveyard by the Sea', trans. C. Day Lewis, from 'Two Translations' from *Collected Poems 1954* by C. Day Lewis (Hogarth Press). Reprinted by permission of Jonathan Cape Ltd., for the Executors of the Estate of C. Day Lewis.

Arthur Waley: 'No discharge' from *The Secret History of the Mongols* (1963). On Chuang Tzu from *Three Ways of Thought in Ancient China* (1939). 'Fighting south of the ramparts' from *Chinese Poems*, trans. Arthur Waley (1946). Reprinted by permission of George Allen & Unwin Ltd.

Sylvia Townsend Warner: 'Graveyard in Norfolk' from *Collected Poems*. Copyright © 1982 by Susanna Pinney and William Maxwell, Executors of the Estate of Sylvia Townsend Warner. Reprinted by permission of Carcanet New Press Ltd.

Evelyn Waugh: from 'Half in Love with Easeful Death' from *A Little Order* (Eyre Methuen). Reprinted by permission of A. D. Peters & Co. Ltd.

Franz Werfel: from *The Forty Days of Musa Dagh*, trans. Geoffrey Dunlop (Hutchinson, 1945). Reprinted by permission of Mrs Anna Mahler.

Edith Wharton: from *A Backward Glance*. Copyright 1933, 1934 by William R. Tyler. © renewed 1961, 1962. Reprinted by permission of A. Watkins Inc. and Laurence Pollinger Ltd. for the Estate of Edith Wharton.

John Hall Wheelock: 'Earth' in *The Gardener and Other Poems*. Copyright 1961 John Hall Wheelock. Reprinted with permission of Charles Scribners' Sons.

Anna Wickham: 'The Free Intelligence' from *Selected Poems*. Reprinted by permission of Chatto & Windus Ltd.

Richard Wilbur: 'The Pardon', from *Poems 1943–56*, and also in *Ceremony and Other Poems*, copyright 1950, 1978 by Richard Wilbur. Reprinted by permission of Faber & Faber Ltd. and Harcourt Brace Jovanovich Inc.

Lyall Wilkes: 'Nightmare'. Reprinted by permission of the author.

William Carlos Williams: from 'At Kenneth Burke's Place' from *Collected Later Poems*. Copyright © 1963 The Estate of William Carlos Williams. Reprinted by permission of New Directions Publ. Corp.

Richard Winston: from Heinrich Mann's *Ein Zeitalter wird besichtigt*, trans. Richard Winston in *Thomas Mann: the Making of an Artist*. Reprinted by permission of Constable Publishers and Alfred A. Knopf Inc.

Ludwig Wittgenstein: from *Tractatus Logico-Philosophicus*, trans. C. K. Ogden. Reprinted by permission of Routledge & Kegan Paul Ltd. and Humanities Press Inc.

James Woodforde: from *The Diary of a Country Parson 1758–1802*, ed. John Beresford (1924). Reprinted by permission of Oxford University Press.

Virginia Woolf: from *The Diary of Virginia Woolf* (1926), vol. iii, ed. Anne Olivier Bell; from *The Letters of Virginia Woolf*, vol. vi, edd. Nigel Nicolson and Joanne Trautmann. Reprinted by permission of The Hogarth Press Ltd., the Author's Literary Estate and Harcourt Brace Jovanovich Inc.

W. B. Yeats: from 'The Tower', copyright 1928 by Macmillan Publ. Co. Inc., renewed 1956 by Georgie Yeats; 'An Irish Airman foresees his Death', copyright 1919 by Macmillan Publ. Co. Inc., renewed 1947 by Bertha Georgie Yeats, both from *Collected Poems*. Reprinted by permission of Michael and Anne Yeats, Macmillan, London, Ltd., and Macmillan Publ. Co. Inc.

Andrew Young: 'The Dead Crab' from *Complete Poems*. Reprinted by permission of Martin Secker & Warburg Ltd.

G. M. Young: 'Dialogue with the Dead' from *Epigrams of Callimachus*, trans. G. M. Young (1934). Reprinted by permission of Oxford University Press.

Yuan Mei: from 'The Will of Yuan Mei' from *Yuan Mei: Eighteenth Century Chinese Poet*, trans. Arthur Waley. Reprinted by permission of George Allen & Unwin Ltd. and Stanford University Press.

Letters to *The Times* are reprinted by kind permission of the writers.

While every effort has been made to secure permission, we may have failed in a few cases to trace the copyright holder. We apologize for any apparent negligence.

INDEX OF AUTHORS

INDEX OF UNASCRIBED PASSAGES